HUMAN EVOLUTIONARY GENESIS IS THE KEY TO HUMAN SUPERIORITY

The Meteoric Rise of Humanity on A Distant Celestial Blue Plannet

Elvis Newman

Know the real cause of your Post Traumatic Slave Syndrome
Deny Ignorance, Propaganda and Invisible Prison Cell
Deny deception of Corpenican Proportions

Published in the United States of America

ISBN 979-8-89395-736-5 (SC)
ISBN 979-8-89395-003-8 (HC)
ISBN 979-8-89395-735-8 (Ebook)

Library of Congress Control Number: 2024922450

Newman Publishing LLC
222 West 6th Street
Suite 400, San Pedro, CA, 90731
www.stellarliterary.com

Ordering Information and Rights Permission:

Quantity sales. Special discounts might be available on quantity purchases by corporations, associations, and others. For details, contact the publisher at the address above.

For Book Rights Adaptation and other Rights Permission. Call us at toll-free 1-888-945-8513 or send us an email at admin@stellarliterary.com.

The ongoing study of Sumerian clay tablets continues to captivate various fields such as science, history, archaeology, and cosmology due to its interdisciplinary impact, bridging gaps between seemingly unrelated areas of knowledge. The beauty of Zecharia Sitchin's theory, regardless of its controversial nature, lies in its ability to inspire curiosity and creativity among people, scientists, and scholars. It encourages the development of new theories and the pursuit of discoveries, even in the face of incomplete or unaccepted evidence by the mainstream scientific community or the general populace.

Sitchin's work exemplifies the power of unconventional ideas to spark dialogue and exploration across diverse disciplines. This openness to new interpretations and possibilities fosters a dynamic environment where established boundaries can be challenged, leading to potential breakthroughs in our understanding of history and the cosmos. The theory's enduring influence underscores the importance of intellectual courage and the willingness to explore the unknown, paving the way for future insights that may one day be widely accepted or lead to entirely new paradigms.

"Once you eliminate the impossible, whatever remains, no matter how improbable, must be the truth."

-Arthur Conan Doyle, Sr. quotes

The Guide to our future from our past

"You can't connect the dots looking forward; you can only connect them looking backwards. So you have to trust that the dots will somehow connect in your future. You have to trust in something — your gut, destiny, life, karma, whatever. This approach has never let me down, and it has made all the difference in my life." [Steve Jobs Stanford commencement speech, June 2005]

Dedication

To those shaping the future of our world:

1. The experts confronting the population crisis and championing Total Fertility Rate and New Eugenics.

2. The lobbyists and lawmakers working to advance policies on Fertility Treatment Options and Assisted Reproductive Technologies.

3. The healthcare professionals advocating for the inclusion of In Vitro Fertilization (IVF), Fertility Treatment Options (FTO), and Assisted Reproductive Technologies (ART) in Universal Health Care and insurance plans.

4. The parents who turn to IVF to bring new life into their homes and build their families.

Because:

"The Child is the father of man." – William Wordsworth

"Children are the future of humanity." – Francis Duggan

"The children of today are the leaders of tomorrow." – Nelson Mandela

Preface

THE
ONCE & FUTURE
TEST TUBE
SPECIES
Numbers Of
Test Tube Babies
Born Each Year
200 MILION
TEST TUBES
BABIES IN
THE U.S
90,000
TEST TUBES
BABIES IN
THE U.S
FIRST TEST TUBE
BABY
1978
2024
2100
2200
Years

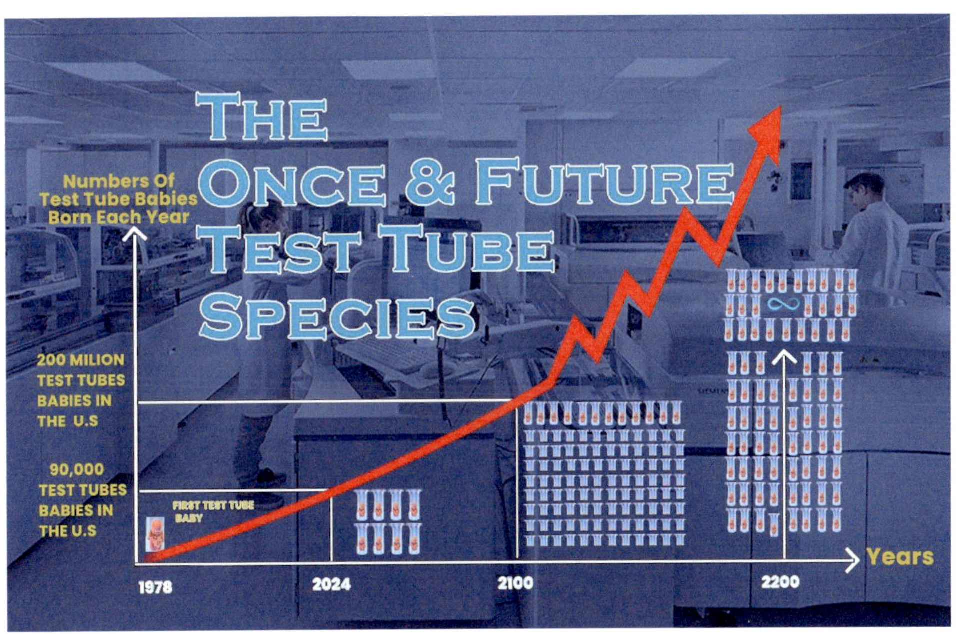

Louise Joy Brown (born 25 July 1978) is an English woman who was the first human to have been born after conception by in vitro fertilization experiment (IVF). In 2021, according to the Centers for Disease Control and Prevention (CDC), approximately 2% of all babies born in the U.S. annually are conceived through IVF and other assisted reproductive technologies. Talk to a group of 100 people born in the United States, and chances are two of them were born as the result of in vitro fertilization, said Dr. Zev Williams, director of the Columbia University Fertility Center.

By 2100, as many as 200 million babies are expected to be conceived using in-vitro fertilization. The notion that humans are somehow a "test tube species" is a speculative and philosophical idea. It may reflect a broader curiosity about human origins and our relationship with technology. In this day and age, the structured environment and access to a broad spectrum of medical resources and expertise in a hospital play a significant role in ensuring the safety and well-being of both the mother and the baby during childbirth.

Things such as Emergency Care, Specialized Staff, Advanced Equipment, Pain Management, 24/7 Availability, Postpartum Care are never available to the animal mothers. Yes, women can give birth alone like the animal mothers, but it's relatively rare and can be very dangerous. The history of midwifery is as old as childbirth itself. Humans are fundamentally different from other species in terms of childbirth and origins. Humans have never been a part of the natural world and have never evolved naturally.

The existence of advanced medical technologies and midwifery from the beginning of human history suggest a deviation from being a natural species from the start. No book, no teacher, no religion, no philosophy, no scientific discovery will ever satisfy our quests for the truths. This book is about the major hidden C.S.I. X file case on human origins that begs to be declassified. Truth always leaves a trace. Hard evidences abound in the world. Let us examine the motives and hidden agendas together.

Contents

Dedication iv

Preface v

Introduction xii

Part 1 The fate of humanity now relies on those, who dare to challenge the future of Earth, by redefining our present, through reexamining our past. 1

Chapter One: Its customary for absurd ideas and fringe sciences to become main stream 4

Chapter Two: Our newly acquired emerging thought process: Quantum Reverse Critical Thinking 7

Chapter Three: Unraveling the Prohibited Secrets of Our Incredible Tech based Modern World through C.S.I. Crime Scene Investigation Principles and Methodology 9

Chapter Four: The Assured Triumph of Human Inevitability and Superiority 13

Part 2: The Once and Future Global Civilization 14

Chapter One: The Cuneiform Clay Tablets Found in Sumeria 16

Chapter Two: Egyptians Pyramids, the Sphinx and Global Megaliths . 18

Chapter Three: Australia's Ancient Egyptian Hieroglyphs 21

Chapter Four: 200,000 B.C.E Metropolis Found in South Africa 23

Chapter Five: Gobekli Tepe, Turkey 25

Chapter Six: Armenia's Carahunge Stone Circle Complex 26

Chapter Seven: The Dogon of Mali, West Africa 27

Part 3: Truths and Misconceptions 29

Chapter One: Mankind Deceived Not Only in Times of Galileo Galilei. The Same Deception Continues . . . Even Today. 31

Chapter Two: Mt. Chimborazo in Ecuador Is Higher Than Mt. Everest by 2,150 Meters / 7,054 Feet 36

Chapter Three: Sizes of China, India, Mexico and Africa Are Grossly Wrong on Google World Map Based on Mercator Projection 39

Chapter Four: Sun Worshipping Cultures Do NOT Worship Our Sun . 43

Chapter Five: "X-Men" Superhumans Could Become a Reality in Thirty Years Say MoD Experts........ 44

Chapter Six: Eisenhower's Greada Treaty with the Aliens in 1954 45

Chapter Seven: Nature's Favorite Numbers— Fibonacci Numbers...... 46

Chapter Eight: Sonic Levitation, Megaliths, and Pyramids, City of Jericho Destroyed by Sonic Vibrations 51

Chapter Nine: The Electric Universe Theory 56

Chapter Ten: Directed Panspermia Proposed by Francis Crick............ 60

Part 4: Be Skeptical Even of One's Own Skepticism........ 62

Chapter One: Map of Antartica Existed, Three Years Before Its Discovery .. 64

Chapter Two: Pre-Colombian Space Shuttle Models and Ancient Indian Flying Crafts 66

Chapter Three: Sunken Cities of the Caribbean, France, India, and Japan 68

Chapter Four: Lessons for Japan and Russia from Ancient Nuclear Reactors........ 70

Chapter Five: A. 7-Year-Old Russian Boy Educated People on Astronomy, Mars Landscape and How to Create High Tech Spacecraft........73

Chapter Six: Irrefutability of DNA Testing as 99.99% Accurate Shattered by Existences of Natural Human Chimeras 77

Chapter Seven: 2000-Year-Old Analog Computer Found........ 79

Chapter Eight: How did the Moon Form? Who Built the Moon? 83

Part 5: New Mentality Brings About New Discoveries........ 87

Chapter One: Monoatomic Gold as Superconductor 88

Chapter Two: Anticipation of Discovery in Astronomy Always Leads to New Discoveries 90

Chapter Three: God Particle Having Been in Existence for Millenia, Found in July 2012 92

Chapter Four: Galileo, Your Best Bet Is to Be Born after Newton 94

Chapter Five: The Greeks Knew the Earth Was Round and Measured Its Circumference Before Copernicus and Galileo 96

Chapter Six: "Fellow Roman Citizens . . . Rome Is Not the Only Mighty Empire in the World!" 99

Chapter Seven: Conversation with the Wright Brothers in 1902, Just before Their Important Breakthrough 100

Chapter Eight: Suppressed Discoveries Become Forbidden and Then Forgotten 101

Chapter Nine: The Elephant of Reality: God, Evolution, and Intelligent Design 104

Part 6: Pursue Truth Aggressively—Essentially and Especially your own Truth! 107

Part 7: What we used to believe 110

Chapter One: Inertia, Indifference and Ignorance 111

Chapter Two: Conversation with the Wright Brothers in 1902 114

Chapter Three: Physical Center of the Universe and Biological Center of the Universe 115

Chapter Four: History Is Wrong 118

Chapter Five: The fates of Human Societies 119

Part 8: Pseudo-Science, Proto Science, True Science 121

Chapter One: Alchemy and The Periodic Table 122

Chapter Two: Discovery of Neptune and Pluto 124

Chapter Three: Faulty Maps to India led Christopher Columbus to the New Continent 126

Chapter Four: Is it better for Galilleo to be born after Newton? 129

Chapter Five: Questions for Roman Citizens 131

Chapter Six: Humans as hybrids of Homo Erectus and Annunaki 132

Chapter Seven: abhorrent types of experimental hominids 134

Chapter Eight: Genesis Revisited: A Scientific Creation Story December 2001 Michael Shermer 136

Chapter Nine: Difficulties on Darwin's Theory 139

Chapter Ten: Sitchin showed us the way The Theory of Human Evolution is Wrong, False, and Impossible 143

Part 9: Sitchin Critics 148

Chapter One: Michael Heiser 149

Chapter Two: Phil Plait 152

Chapter Three: Rob Hafernik 154

Chapter Four: Ian Lawton 155

Part 10: Sitchin Supporters 156

Chapter One: Michael Tellinger 157

Chapter Two: Marshall Klarfeld 159

Chapter Three: Lloyd Pye 161

Chapter Four: Crop Circle Evidences relating to Nibiru 163

Part 11: A brand new interpretation with 21st Century technology vocabularies and scenario descriptions as well as a thorough readaption of the work "The Lost Book of Enki" by Zecharia Sitchin 166

Chapter One: The debate for Annunaki and Ape Hybrids 167

Chapter Two: Story of Garden of Edin, Prehistoric Iraq 169

Chapter Three: Terran Wild Life Natural Reserve and Research Center Garden of Edin, Prehistoric Iraq 170

Chapter Four: A.C. T. G. and adenine, cytosine, thymine and guanine Annunaki Center for Terran Genetics, Garden of Edin, Prehistoric Iraq 172

Chapter Five: Natural Cycle In vitro fertilisation (IVF) 173

Chapter Six: Nimah's Progress Entry 175

Chapter Seven: Ningishzidda's Junk DNA and Abnormalies Another try. Another deformed creature was made. 177

Chapter Eight: Curiosity kills the Annunaki 178
Chapter Nine: FOXP2 language Gene Terran Experiment No. 478,911 . 179
Chapter Ten: Frankenstein in Annunaki Womb 180
Chapter Eleven: Human Prototype Adamu 001 182
Chapter Twelve: Mass Production of Adamu 184
Chapter Thirteen: Wanted! Gestation Surrogate Mothers! Sumerian Africa News Network 185
Chapter Fourteen: Nature's Curse for Hybrid Animals 187
Chapter Fifteen: Genetic Experiment that Goes Horribly Right 188
Chapter Sixteen: ANNUNAKI CLASSIFIED DOCUMENT-- "Genesis" 189
Chapter Seventeen: Expulsion from Garden of Edin 191
Chapter Eighteen: The Descent of Man 192
Chapter Nineteen: Man's First Society 193
Chapter Twenty: Homo Sapien's final genetic make over 194
Chapter Twenty-One: Annunaki Xenobiologists' Science Digest 195
Chapter Twenty-Two: Annunaki confronts changing demographics Sumerian Africa News Network 196
Chapter Twenty-Three: Global Flood for Humanity's destruction 197
Chapter Twenty-Four: Lording Over the Humans 198
Chapter Twenty-Five: The quest for Global Domination 199
Chapter Twenty-Six: Prehistoric WW3 erupted 200

Part 12: Enki's Lessons for Humanity 201

Chapter One: Enki's open letter to humanity: Avoidance of last World War 202
Chapter Two: Last World War in Sumer 203
Chapter Three: Memoirs of A Cockroach 2024 B.C. 206
Chapter Four: We are waiting patiently for you to join us in the wonderful Galactic Journey of Exploration 207

Part 13: Conclusion 208

Introduction

It is my sincere hope that everyone who reads this work will be inspired to question things and to search out these and other new truths and discoveries for themselves. We should all be engaged in the most important intellectual dialog of exchange and sharing for the twenty-first century.

An Invitation to Seek Truth:

In a world where information is abundant, but true understanding often remains elusive, it is my sincere hope that everyone who reads this work will be inspired to question the status quo, to seek out new truths, and to discover the unknown for themselves. This book is not meant to be a definitive answer, but rather a starting point for a deeper exploration of the mysteries that have shaped our world and our perception of it. We live in an age where intellectual curiosity and critical thinking are more important than ever. The 21st century demands that we engage in the most important intellectual dialogue, exchange, and sharing of ideas. It is through this process that we can hope to uncover new insights and build a more enlightened society.

A Call for Independent Investigation:

I do not ask, nor do I expect, anyone to blindly believe what is written within these pages. Instead, I encourage you to investigate all the evidence for yourself. In the quest for truth, it is essential that we approach the world with an open mind, willing to challenge our preconceived notions and explore new perspectives. As we do this, we will find that our shutters will be opened onto a beautiful landscape—one that reveals life and society in all its variety, with an ever-greater sensitivity and realization. This journey of discovery is not just about accumulating knowledge, but about developing a deeper understanding of the world around us and our place within it.

The Nature of Truth:

Truth is a concept that is often elusive and complex. It can be stranger than fiction, sometimes vastly outnumbered, and often difficult to discern amid the noise of misinformation and preconceived biases. Truth may lie out there somewhere, waiting to be uncovered through diligent research and open-minded exploration. Yet, it may also lie within us, in our inner awakening, in the realization that our understanding of the world is constantly evolving. The pursuit of truth requires courage, humility, and the willingness to embrace uncertainty. It is not a destination, but a journey—a continuous process of questioning, learning, and growing.

The Unifying Power of Discovery:

One of the central themes of this work is the exploration of the Sumerian clay tablets, a discovery that has sparked great controversy and debate. It is my hope that the interchanges and interactions surrounding this discovery will not divide us, but rather unite us in a shared quest for understanding. In a world that often feels fragmented and polarized, the search for truth has the power to bring us together in a beautiful harmony called love. This unity is not based on uniformity of thought, but on a shared commitment to seeking out the truth, no matter where it leads. In our quest for the ultimate truth, we have the potential to usher in a new era—a Golden Age characterized by wisdom, compassion, and a deeper connection to one another.

A Thought-Provoking Journey:

As you read this book, I thank you for taking the time to engage with the ideas presented here. My hope is that this work will serve as a thought-provoking catalyst for your own journey of discovery. Use it as a tool to inspire your curiosity, to challenge your assumptions, and to explore the vast landscape of knowledge that awaits you. In the end, the true value of this book lies not in the conclusions it offers, but in the questions, it raises and the conversations it sparks.

The quest for truth is a noble endeavor, one that requires us to look beyond the surface and to seek out deeper meanings and connections. It is a journey that can transform not only our understanding of the world, but also our understanding of ourselves. As we embark on this journey together, let us do so with open hearts and open minds, ready to embrace the challenges and rewards that come with the pursuit of knowledge and wisdom.

Part 1

The fate of humanity now relies on those, who dare to challenge the future of Earth, by redefining our present, through reexamining our past.

In the 21st century, the rapid rise of technologies like Artificial Intelligence (AI), Virtual Reality (VR), and Genetic Engineering has not only transformed the future of humanity but also fundamentally altered our understanding of identity and existence.

1. AI and the Redefinition of Thought:

Machine Learning and Human Cognition: AI has changed how we process information and make decisions. Algorithms increasingly influence our thinking, prompting us to question the nature of intelligence and consciousness. As machines take on tasks that require problem-solving and pattern recognition, the boundaries between human thought and machine processing blur, challenging the uniqueness of human cognition.

2. Virtual Reality and the Nature of Experience:

Merging Physical and Virtual Worlds: VR creates immersive environments that can feel as real as the physical world. This convergence of the virtual and physical prompts a reevaluation of what constitutes "reality." If experiences in a simulated world can evoke genuine emotions and responses, what does this say about the nature of existence? How do we distinguish between the "real" self and the avatars we inhabit in virtual spaces?

3. Genetic Engineering and the Rewriting of Life:

Modification of Biology: With advancements in genetic engineering, we can now alter the fundamental building blocks of life. This power to modify our genes—whether to eliminate diseases, enhance physical traits, or even extend life—forces us to reconsider what it means to be human. The line between natural evolution and artificial intervention is becoming increasingly blurred, raising ethical questions about identity and the essence of our biological selves.

4. Redefining Self-Awareness and Existence:

Descartes Revisited: Descartes' declaration, "I think, therefore I am," has long served as a cornerstone of self-awareness and existence. However, in an age where machines think and virtual identities thrive, this statement demands reevaluation. What does it mean to "think" in a world where AI can simulate thought processes? How does existence manifest in digital realms? The traditional understanding of self-awareness is being reshaped, calling for a new philosophical framework that accounts for these technological realities.

In this context, the integration of AI, VR, and genetic engineering into our lives compels us to redefine what it means to be human. These technologies are not just tools; they are catalysts for a profound transformation in how we perceive ourselves and our place in the world. As we navigate this new era, the challenge lies in understanding and embracing these changes while maintaining a sense of identity and purpose.

Chapter One
Its customary for absurd ideas and fringe sciences to become main stream

As technologies like Artificial Intelligence (AI), Virtual Reality (VR), and Genetic Engineering transition from fringe science to mainstream applications, they challenge long-held concepts of human cognition, identity, and reality. These technologies, once considered speculative or even outlandish, now push us to reconsider fundamental ideas like Descartes' "I think, therefore I am" (cogito ergo sum). In the face of these advancements, there is an urgent need to redefine self-awareness and existence, as the lines between human and machine, reality and virtuality, become increasingly blurred.

Historical Precedents: Fringe to Foundation

The shock lies in realizing that many ideas now central to our understanding of the world were once considered fringe theories. Just as these groundbreaking ideas eventually became pillars of modern science, so too could today's emerging technologies reshape our future understanding of existence and identity.

1. Germ Theory:

From Skepticism to Foundation: Initially met with resistance, the germ theory, developed by Louis Pasteur and Robert Koch in the 19th century, revolutionized medicine and public health. It shifted the understanding of disease from miasma (bad air) to microorganisms, laying the groundwork for modern hygiene, antibiotics, and vaccines.

2. Plate Tectonics:

Controversy to Cornerstone: Once a contentious idea, plate tectonics became the bedrock of geology by the mid-20th century. It explains the movement of Earth's lithospheric plates, providing crucial insights into natural phenomena such as earthquakes, volcanic activity, and the formation of mountains.

3. Heliocentrism:

Revolutionary to Reality: Nicolaus Copernicus' 16th-century proposal that the Earth orbits the Sun faced initial rejection but ultimately upended the geocentric model of the universe. This shift not only revolutionized astronomy but also altered humanity's place in the cosmos, paving the way for modern science.

4. Quantum Mechanics:

Fringe to Fundamental: In the early 20th century, quantum mechanics challenged classical physics with its strange, probabilistic nature. Today, it is essential to our understanding of the atomic and subatomic world, underpinning technologies like semiconductors, lasers, and quantum computing.

5. Relativity:

Revolutionary to Integral: Albert Einstein's theories of Special and General Relativity, which were once revolutionary and difficult to accept, are now integral to modern physics. They explain the nature of space, time, and gravity and have practical applications in technologies like GPS.

6. Natural Selection:

Controversial to Core Principle: Charles Darwin's theory of natural selection, introduced in the mid-19th century, was initially met with fierce opposition. Today, it is a cornerstone of evolutionary biology, explaining the diversity of life on Earth and the process of adaptation.

7. Big Bang Theory:

Speculation to Standard: Proposed in the 20th century as a theory for the origin of the universe, the Big Bang Theory is now the leading explanation for the universe's development. Supported by evidence such as cosmic microwave background radiation and the expansion of the universe, it shapes our understanding of cosmology.

Conclusion:

The journey from skepticism to acceptance seen in these historical examples highlights the transformative power of ideas once on the fringes of science. As AI, VR, and genetic engineering continue to evolve, they could similarly redefine our understanding of self-awareness, existence, and reality. Just as past revolutionary theories reshaped the foundations of science, today's emerging technologies may lead to new paradigms that challenge and ultimately expand the boundaries of human knowledge and experience.

Chapter Two

Our newly acquired emerging thought process: Quantum Reverse Critical Thinking

Our newly acquired emerging thought process is a reflection of the profound changes brought about by AI, VR, genetic engineering, and other cutting-edge technologies. It is a blend of human intuition and machine efficiency, physical reality and virtual experiences, biological knowledge and ethical considerations. This evolving thought process is reshaping our understanding of the world and ourselves, pushing us to redefine concepts of identity, intelligence, and existence in an increasingly complex and interconnected reality.

Quantum Thinking: Quantum thinking often refers to the application of principles from quantum physics to problem-solving and creativity. It emphasizes non-linear, probabilistic, and interconnected approaches rather than traditional linear and deterministic ones. Quantum thinking encourages exploring multiple possibilities and perspectives simultaneously.

Reverse Critical Thinking: Reverse critical thinking involves analyzing a problem or concept from the opposite direction or challenging conventional assumptions. Instead of critically evaluating something from the established norms or traditional methods, it means approaching it from an unconventional angle to uncover new insights or solutions.

Quantum Reverse Critical Thinking will help us in:

Exploring Multiple Possibilities: Like quantum thinking, this approach would involve considering various potential outcomes and solutions, not just sticking to one fixed path or solution.

Challenging Conventional Assumptions: This aspect would focus on questioning and reversing traditional assumptions and norms related to the problem at hand.

Embracing Uncertainty and Complexity: It would embrace the idea that multiple realities or solutions might coexist, reflecting the probabilistic nature of quantum theory.

Non-linear Problem-Solving: Using non-traditional methods to approach problems, potentially finding innovative solutions that wouldn't emerge from linear thinking alone.

In essence, quantum reverse critical thinking could be a method of problem-solving that combines the flexibility and multi-faceted nature of quantum thinking with the innovative, boundary-pushing aspects of reverse critical thinking. It’s about exploring possibilities in a non-linear fashion while challenging the conventional wisdom that might limit creative solutions.

Chapter Three
Unraveling the Prohibited Secrets of Our Incredible Tech based Modern World through C.S.I. Crime Scene Investigation Principles and Methodology

Understanding historical contexts provides clarity on how past actions and trends impact our current world, guiding us in making informed decisions and preparing for the future. By learning from history, we can better navigate today's complexities and shape tomorrow's possibilities. A deeper understanding of our past involves examining historical events, cultural shifts, and societal developments to uncover underlying patterns and influences. This insight helps us recognize how past decisions and trends have shaped the present, offering valuable lessons for guiding future choices. By analyzing these historical contexts, we can gain perspective on current issues and better anticipate future challenges and opportunities.

Physical evidence for ancient civilizations comes from a variety of archaeological and geological sources. Here are some key types of evidence:

Archaeological Artifacts:

Tools and Weapons: Items like stone tools, metal weapons, and pottery shards reveal technological advancements and daily life.

Art and Inscription: Carvings, paintings, and written records (like clay tablets and papyri) provide insights into cultural, religious, and political aspects.

Architectural Remains:

Buildings and Structures: Ruins of temples, palaces, and residential areas showcase architectural styles and urban planning.

Monuments: Large-scale constructions like pyramids, ziggurats, and stone circles (e.g., Stonehenge) reflect the engineering skills and societal priorities.

Settlement Patterns:

City Layouts: Excavations of ancient cities reveal streets, houses, and public spaces, illustrating the organization of communities.

Infrastructure: Remains of roads, aqueducts, and irrigation systems highlight advancements in engineering and resource management.

Burial Sites and Graves:

Tombs and Cemeteries: Burial sites, including graves and burial goods, offer information on social hierarchies, rituals, and customs.

Mummies and Skeletons: Physical remains provide data on health, diet, and lifestyle.

Pottery and Ceramics:

Vessels and Utensils: Pottery shards and complete vessels reveal trade practices, artistic styles, and daily activities.

Written Records:

Documents and Texts: Ancient manuscripts, legal codes, and administrative records provide direct insights into governance, economics, and culture.

Environmental Evidence:

Pollen and Soil Samples: Analysis of ancient pollen and soil layers helps reconstruct past climates and agricultural practices.

Animal and Plant Remains: Faunal and floral remains indicate diet and domestication practices.

Numismatic Evidence:

Coins and Currency: Coins and other forms of currency offer clues about trade, economy, and political systems.

These physical remnants collectively help historians and archaeologists piece together the lives, achievements, and societies of ancient civilizations.

Applying CSI (Crime Scene Investigation) methodology to ancient civilizations involves adapting forensic techniques to archaeological contexts. Here's how these principles can be used:

Evidence Collection:

Systematic Excavation: Careful, methodical digging to uncover artifacts and structures while preserving context.

Stratigraphy: Analyzing layers of soil and sediment to understand chronological sequence.

Preservation and Documentation:

Proper Handling: Using tools and methods to avoid damage to artifacts and remains.

Detailed Records: Keeping accurate records of locations, conditions, and context of finds.

Analysis:

Forensic Examination: Analyzing physical evidence (like bones, tools, and pottery) to interpret past human activities and conditions.

Laboratory Techniques: Employing methods such as radiocarbon dating, DNA analysis, and isotopic studies for precise dating and understanding.

Reconstruction:

Site Reconstruction: Rebuilding ancient structures or layouts based on collected evidence.

Behavioral Reconstruction: Inferring daily life, social structures, and events from material remains.

Interdisciplinary Approach:

Collaboration: Working with experts in various fields (e.g., anthropology, geology, history) for comprehensive analysis.

Technology Use: Utilizing tools like GIS (Geographic Information Systems) and 3D modeling for detailed analysis and visualization.

Ethical Considerations:

Respect for Cultural Heritage: Ensuring sensitive treatment of artifacts and remains, and respecting the cultural significance of findings.

These methods help uncover, preserve, and understand the complexities of ancient civilizations.

Just as the extraordinary story of human civilization redefined our place in the universe, today's technological revolution is reshaping our understanding of what it means to be “Human” in the 21st century.

Chapter Four
The Assured Triumph of Human Inevitability and Superiority

The modern world as we know it is the result of a complex and multifaceted historical process. Each of the elements represents a piece of the puzzle, coming together over centuries to create the modern world. It's a narrative of continual change, influenced by technological progress, political upheavals, economic transformations, and cultural shifts.

In the following Parts and Chapters of this book, the author will:

Review the Evidence: Examine all the collected data—artifacts, remains, documents, and environmental samples.

Analyze the Findings: Interpret the evidence to understand its significance. For ancient civilizations, this might include analyzing artifacts, burial practices, and environmental conditions.

Develop Hypotheses: Formulate possible scenarios based on the evidence. Consider how these scenarios fit with known historical and cultural contexts.

Construct the Narrative: Piece together the most plausible sequence of events. Describe the social, economic, and environmental conditions that influenced these events.

Validate with Context: Cross-check the narrative with historical records and other archaeological findings to ensure accuracy and consistency.

Present Findings: Share the reconstructed story through reports, presentations, or publications, highlighting key insights and their implications for understanding the civilization.

By following these steps, the author aims to craft a detailed, informed, and incredible story that not only connects past events and societies but also explains how these historical developments and emerging technologies have shaped the modern world. This narrative will provide readers with a clearer and more realistic understanding of the complexities that have brought us to where we are today, illustrating the interconnectedness of past and present in creating our incredible modern society.

Part 2
The Once and Future Global Civilization

"All things are wearisome, more than one can say.

The eye never has enough of seeing, nor the ear its fill of hearing.

What has been will be again, what has been done will be done again; there is nothing new under the sun.

Is there anything of which one can say,

Look! This is something new?

It was here already, long ago; it was here before our time.

No one remembers the former generations, and even those yet to com will not be remembered by those who follow them."

--King Solomon

Just when we pride ourselves to have come very far to represent the apex of the world's civilizations, having the best forms of democracy, social institutions and the leading edge in technology, we are learning more and more about a prehistoric advanced societies that was planet wide and also highly technical. What we think of as our earliest societies, the Sumerians Egyptians, Babylonians, Greeks, Romans, Indians and Chinese are nothing but the remnants of this earlier prehistoric global civilization. The evidence is everywhere. Avalanche of evidences exploded every now and then in new archaeological finds.

Some 445,000 years ago, Enki, the Most Renowned Genetic Scientist from Planet Nibiru discovered that Erectus-Neanderthal genome could assist acclimatization of Nibiran Genome to Earth ecosystem and environment by the process of genetic hybridization. The successful colonization of Earth as a Mining Operation would be feasible with the creation of a slave species: Homo sapiens. Enki was subsequently mistaken and worshipped as The God by many cultures throughout the world.

Our religions were established by these Anunnakis. Myths, fallacies and wrongful adaptations were passed down from generations after generations. The only evidence on the "Existence" of Jesus Christ would have to be a verifiably contemporary eyewitness account by someone other than an avowed disciple or follower of Jesus (like, say, the chef at the Last Supper, assuming he wasn't Christian). Other than the Bible, Jesus Christ should also be mentioned in many historical writings of his time, or shortly thereafter. There should be voluminous mentioning and cross referencing of Jesus in many of the contemporary books and media of his day. Look at the publicity we have given our celebrities, look at the Media coverage of Prince William and Princess Kate around the globe...

The only other non-biblical sources, who wrote their very brief references to Christ at least a century after his purported crucifixion, are those of Josephus and Tacitus. Furthermore, even in the Gospels, there are many inconsistencies. There are many years of unrecorded and unaccounted for life in Jesus's upbringing. The fact is that, until all such hard evidences are produced, Jesus Christ must remain a terrifically inspirational but entirely faith-based, non-historical, fictional character.

The key to our future lies in our understanding of our past...

Chapter One

The Cuneiform Clay Tablets Found in Sumeria

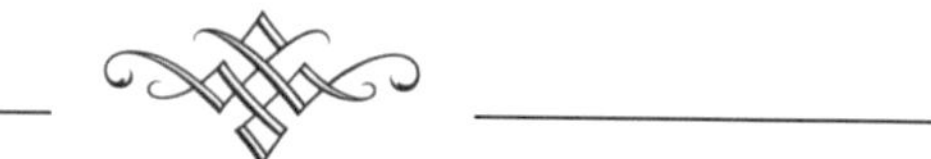

Cuneiform writings had been saved on tens of thousands of clay tablets that have been found in Mesopotamia, in the region of modern Iraq, in the last one hundred and fifty years. Through clay tablets, cylinder seals and stele, the Sumerians have provided us with a graphic and richly detailed version of man's early histories including the story of creation, both of the Earth and of man.

The Sumerian account, through the paraphrasing of scholar Zecharia Sitchin, is really the only account that provides totally a plausible series of events that adequately explains every single puzzle we are faced with. The fact that the Sumerians knew about Uranus, Neptune, and Pluto way before anyone else raises the question: If their story is untrue, then how on Earth were they able to acquire such accurate and detailed knowledge?

Uranus was discovered by William Herschel on March 13th of 1781. This was the first time a planet was discovered using a telescope. Neptune was discovered on Sept 23rd 1846 . . . and only by math prediction due to Uranus' orbit rather than actual observation. Pluto wasn't discovered until 1930.

It appears obvious that both of the latter accounts of Babylonian and Bible creation stories were heavily influenced by the much longer and more detailed earlier Sumerian story and can be easily confirmed, as many parallels stories can be seen in all of them.

From their texts, it was clear that the Sumerians had quite a significant amount of scientific and astronomical knowledge imparted by their creator, the Anunnaki.

Chapter Two
Egyptians Pyramids, the Sphinx and Global Megaliths

The enigma of pyramids and megalithic sites throughout the world has intrigued archeologists, historians, and engineers alike, and none of them have provided logically satisfying answers to the enigma. The very fact that one to two hundred tons of stone blocks had been cut precisely with laser precision and hauled close to five hundred feet above the ground, a feat impossible even with today's technology, is a strong indication of ancient technological achievement. It is something that we cannot dismiss off hand. And these pyramids and megaliths may even be thousands of years older than what has been thought.

- Pyramid Complex. Sedeinga, Sudan
- *One overlooked fact is that Sudan has more pyramids than Egypt.*
- *Ancient kingdoms of Kush and Nubia once rivaled Egypt, Greece and Rome.*
- Pyramid Hill, Lake Baikal, Russia
- Tenerife Pyramids, Canary Islands
- Baalbek, Lebanon

The massive and elegant Roman stonework and columns pale by comparison to the megaliths they were built upon. The temple very visibly incorporates into its foundation stones of some fifteen hundred tons. They are some 68 x 14 x 14 feet! They are the largest worked stones on Earth! It is a mystery how such stones could have been moved into place, even according to our science and engineering knowledge of today. It is also a fact the Romans did not use this type of stonework.

- Mount Etna Pyramids, Sicily, Italy.
- Ligourion Pyramids, Greece
- Pyramids Complex, Xi'an China
- *The world's largest pyramid is rumored to be in Qin Lin County in a "forbidden zone" of China. Estimated at nearly one thousand feet high and made of impounded Earth and clay, holding vast tombs.*
- Almendres Cromlech, Alentejo, Portugal
- Avebury Henge, Avebury, England
- Machu Picchu, Cusco, Peru.
- *There are many odd shaped carved stones. Why would these ancient people build a city at such a high and treacherous altitude, where the valleys below would provide all their needs? Why are there just a few skeletal remains, most of them women, for such a large complex?*

- Callanish Stones, Isle of Lewis, Scotland
- Anta Grande do Zambujeiro, Alentejo, Portugal
- Easter Island, Chile
- *For centuries, scientists have tried to solve the mystery of how the colossal stone statues of Easter Island moved.*
- Carnac Stones, Carnac, France
- Horca del Inca, Copacabana, Bolivia
- Stonehenge, Wiltshire, England
- *Archaeological evidence reveals that pigs were slaughtered at Stonehenge in December and January, suggesting possible celebrations or rituals at the monument around the winter solstice.*
- America's Stonehenge, Salem, New Hampshire, U.S
- Monk Mound, St. Louis, Illinois, U.S.
- Puma Punku, Tiwanaku. Bolivia
- *Careful examination of the stones revealed intricate stonework, as though machine tools or even lasers were used. Puma Punku is located at an altitude of 12,800 feet. No trees grew in that area. The Ceramic Fuente Magna Bowl has Sumerian cuneiform and proto-Sumerian hieroglyphics written on it. There are several stones at Puma Punku that weigh over one hundred tons.*

Many of these large pyramid complexes and megalithic sites have over time been closely examined by engineers, investigators, and experts. What has been noticed is that there are definite signs of tooling and automation. The precision achieved is definitely not handcrafted and requires very high-grade engineering equipment.

The precision achieved is definitely not handcrafted and requires very high-grade engineering equipment.

The methodical and meticulous Egyptians kept very careful records about everything they ever did, every king they had, every war they fought, and every structure they built. Why didn't they record how the pyramids were built? Purportedly strange new species of fungi were found in King Tutu's chamber. Again, why?

Chapter Three
Australia's Ancient Egyptian Hieroglyphs

Five thousand years old Egyptian hieroglyphs have been found in New South Wales, Australia.

The two hundred and fifty hieroglyphs are certainly not your average aboriginal animal carvings but something clearly alien in the Australian bush setting. They tell the tale of early Egyptian explorers, injured and stranded, in ancient Australia. The discovery centers around a most unusual set of rock carvings found in the National Park Forest of Hunter Valley, one hundred kilometers north of Sydney.

The hieroglyphs were extremely ancient, in the archaic style of the early dynasties.

Egyptologist Ray Johnson claimed that these glyphs stemmed from the third dynasty of Egypt and chronicle a tragic saga of ancient explorers shipwrecked in a strange and hostile land.

Having spent many years learning and living with the Bundjalung and Gumilaroi people of northern New South Wales, authors Steven Strong and Evan Strong, after consultation and investigation with elders of Australia, believed they had rediscovered a hidden history on the origins of humanity in Australia and the world. Their books try to prove through scientific facts that which the elders of Australia insist is true.

Their claim is supported by genes, mtDNA, and blood evidence. There are also many experts who concur with them that aboriginal people set sail from Australia, not to, fifty thousand years ago. They claim that aboriginal people sailed to and settled in America over forty thousand years ago and visited many other places including Egypt, Japan, Africa, India, etc. They were the first Homo sapiens who evolved before the sapiens of Africa and who gave the world art, axes, religion, marine technology, culture, cooperativc living, language, and surgery.

According to the Strongs, "The debate over whether they were the first people in America is virtually a closed case. Hundreds of bones and skulls have been discovered that are undeniably of 'Australian Aboriginal' origin."

Professor Clive Gamble explains, "We have to construct a completely new map of the world, and how it was peopled."

Chapter Four
200,000 B.C.E Metropolis Found in South Africa

About one hundred and fifty miles inland, west of the port of Maputo, South Africa is the remains of a huge metropolis that measures fifteen hundred square miles within an even larger community of ten thousand square miles. It was estimated to be constructed around 200,000 BCE!

Michael Tellinger tells us, "The photographs, artifacts, and evidence we have accumulated points unquestionably to a lost and never before seen civilization that predates all others—not by just a few hundred years, or a few thousand years . . . but many thousands of years. These discoveries are so staggering that they will not be easily digested by the mainstream historical and archaeological fraternity, as we have already experienced. It will require a complete paradigm shift in how we view our human history."

"The thousands of ancient gold mines discovered over the past five hundred years point to a vanished civilization that lived and dug for gold in this part of the world for thousands of years," says Tellinger. "And if this is in fact the cradle of humankind, we may be looking at the activities of the oldest civilization on Earth. We find roads— some extending a hundred miles—that connected the community and terraced agriculture, closely resembling those found in the Inca settlements in Peru."

The individual ruins mostly consist of stone circles. Most have been buried in the sand and are only observable by satellite or aircraft. Some have been exposed when the changing climate has blown the sand away, revealing the walls and foundations.

Chapter Five
Gobekli Tepe, Turkey

Located about five hundred miles away from Istanbul is Sanliurfa, Turkey, where Gobekli Tepe, the oldest human-made place of worship was discovered. The massive sequence of stratification layers suggests several millennia of activity, perhaps reaching back to the Mesolithic.

- The lowest layer contains monolithic pillars linked by coarsely built walls to form circular or oval structures.
- The middle layer has revealed several adjacent rectangular rooms with floors of polished lime.
- The uppermost layer consists of sediment deposited as the result of agricultural activity.

The monoliths are drawn with pictures of lions, bulls, boars, foxes, gazelles, asses, snakes and other reptiles, insects, arachnids, and birds, particularly vultures and water fowl. The T-shape pillars have carved arms to represent anthropomorphic gods.

The German archeological team has taken fifteen years to uncover about five to 7 percent of a gigantic civilization and proven that Gobekli Tepe is almost twelve thousand years old, much older than Egypt, Greek, Roman, Chinese, Indian, and Hebrew history. No other site is this advanced and is this old.

Graham Hancock says that we have this gigantic site with huge circular megalithic structure, that stands there as a mystery, asking us to go figure how was this done, what's the background to this. We have no idea, who made them. They just come out of the dark of the last ice age, about which we know nothing, and enter this stage of history already fully formed. And to his mind this is indicative of a major forgotten episode of human history.

Since the site of Gobekli Tepe is more than twelve thousand years old, it has extended the first human civilization by more than seven- thousand years from Sumer which predates all ancient civilizations still in existence.

Chapter Six
Armenia's Carahunge Stone Circle Complex

A new study into Armenia's Carahunge stone circle complex has shown that it is one of the oldest known megalithic sites, dated around 5500 BC. According to Russian prehistorian professor Paris Herouni, "Carahunge was created as an astronomical observatory marking the movement not only of the sun and moon, but also the stars. Carahunge's principal stellar alignment is towards Deneb, the bright star in the constellation of Cygnus the swan."

According to Andrew Collins, "Herouni ran the angle of the stone through various astronomical programs and found that it was aligned to Deneb at a date of around 5,500 BC, suggesting that this was the time frame in which Carahunge was in use by an advanced society of astronomer priests. It was this alignment that provided the key to finally dating the site, which was expected to have been constructed during this distant epoch."

Chapter Seven
The Dogon of Mali, West Africa

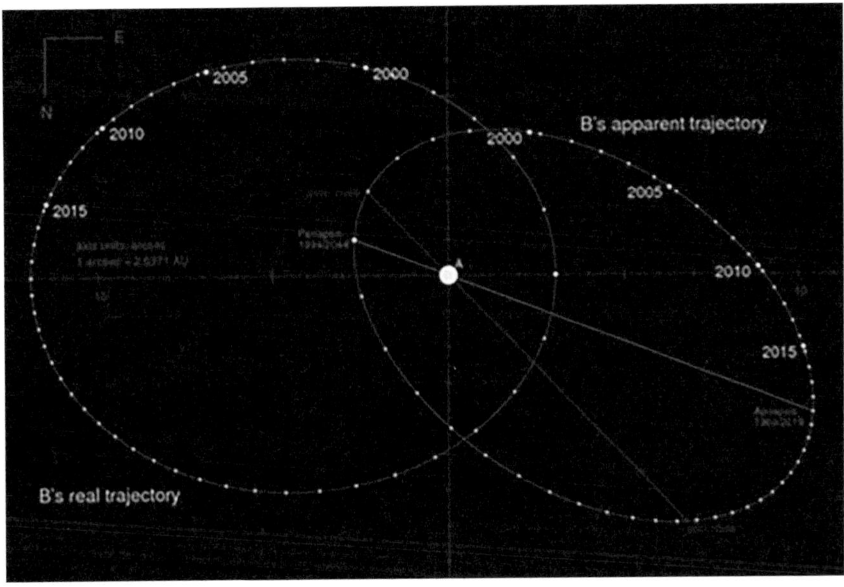

In his book, *The Sirius Mystery,* Robert Temple tells us that in the early twentieth century two French anthropologists named Marcel Griaule and Germain Dieterlen spent a good deal of time living with the Dogon in order to study their ways. In 1930, after they had been living with the tribe for some 15 years, four Dogon Priests decided that it was time to take the Frenchmen into their confidence and invited the men to share in the tribe's most important and sacred tradition. The tale was the secret Dogon creation myths about their sacred star Sirius which located some 8.6 light years from Earth. Sirius is also the brightest star in the night sky. The Dogon told the Anthropologists that Sirius was the home of the Gods who had made them who they are—some eight thousand years ago. They told them that Sirius is the smallest and heaviest thing there is and that it was white in color. They said that it had a companion star, invisible to the human eye but that it moves around Sirius in an elliptical orbit that took fifty years. They said Sirius was incredibly heavy and that it rotated on its axis and they further describe it as having a circle of reddish rays around it that is 'like a spot spreading but staying still.' Dogon oral traditions also quite adamantly state that they have known for thousands of years that Jupiter had moons and Saturn had rings around it.

What they discovered was that the Dogon had in fact, accurately described the three principal properties of a white dwarf star: small, heavy and white and had also stated that Sirius is a binary star, both of which we now know the Sirius system to be. They are also absolutely correct in their knowledge of its companions' rotation as Sirius-B orbits Sirius-A every 49.9 to 50.0 years.

The Dogon people also use an extremely unorthodox calendar that is based on a fifty-year cycle. This cycle is uniquely unusual because it does not follow any cycles coinciding with any movements of our Earth, moon or sun but instead is based wholly on the rotational movements of Sirius B. In fact, the entire Dogon Culture is based around the fifty- year cycle of Sirius B. The Dogon People of pre 1930 had no telescopes or real written language. How is it they were able to accurately describe things we still only possessed a very limited knowledge of? Where did they get their information? The Dogon repeatedly say that they were taught these things many, many years ago by their Gods who visited them from their home planet that orbits Sirius B. The Dogon also describe them as being amphibious creatures.

There happens to be too many uncanny concurrences between what the Dogons saw then and what modern scientists see now, and ones that cannot be shrugged off as being mere coincidences.

Part 3
Truths and Misconceptions

Many years have passed since the 2008 financial crisis struck. And only now do Wall Street executives seem to be gaining awareness that many potential clients have lost trust in them and the stock market, and have been voting with their feet. When the stock market was plunging in 2008, some brokers and advisers hid instead of taking calls from clients.

Should we place absolute faith in our financial, scientific and political leaders?

The conclusion is simple: Distrust the experts.

Why? Because you don't know their incentives, and they can make the models say whatever is politically useful to them. This is a manipulation of the public's trust of mathematics, but it is the norm rather than the exception. And modelers rarely if ever consider the feedback loop and the ramifications of their findings or theories on our culture. The truth is somewhat harder to understand, a lot less palatable, and much more important.

This raises a larger question: how can the public possibly sort through all the noise? Whose job is it to push back against rubbish disguised as authoritative scientific theory? When should we stop trusing the icons?

In Sep 2011, the European Centre for Nuclear Research (CERN) showed that the accepted speed of light could be broken by sixty nanoseconds, or more precisely that neutrinos arrived sixty nanoseconds too early than if they were travelling at the speed of light, breaking the speed of light itself by 0.00248%. Two experiments reached the same conclusion. This is a finding that violated Einstein's venerable theory of special relativity that states that nothing in the universe can travel faster than the speed of light in a vacuum.

The journal Science reported that a faulty connection between a GPS and a computer likely caused the anomaly, thus leaving Einstein's theory intact. Scientific findings are not set in stone, and even at the time, CERN doubted its own findings. They agreed that the finding could be an error, saying that even the tiniest of shifts in the Earth between the two measuring points could have changed the results.

General Relativity predicts a singularity at the center of a black hole - most physicists believe this is wrong but no one can prove that yet. A unification of quantum mechanics and General relativity may resolve this.

General Relativity may be wrong in certain circumstances. Rotations of stars in galaxies don't conform to General Relativity. Scientists have 'invented' dark matter to explain this, but since no one knows what dark matter is, it may be that General relativity is wrong under certain conditions. There have been various attempts to add vector and scalar components to Einstein's equations, but so far, they haven't been consistent with observations.

With man's ever-increasing knowledge of science and technology, ancient, half-forgotten legends seemingly have no place. Truths and ideas unable to be examined using orthodox methodology are automatically scorned and ridiculed.

But what happens to the big picture when what we see and what we think we see in the small picture with our limited minds and observations are two very different things?

What are some of our controversial truths and misconceptions?

Chapter One
Mankind Deceived Not Only in Times of Galileo Galilei. The Same Deception Continues . . . Even Today.

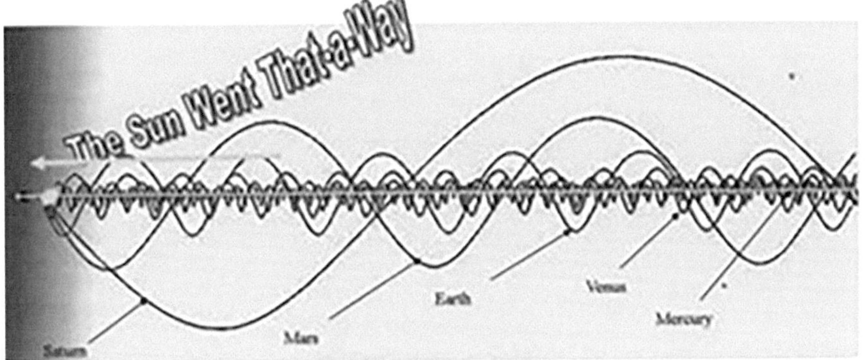

The Planetary Vortex
Movements of the inner planets viewed over a period of one Saturn cycle of 29.36 years.

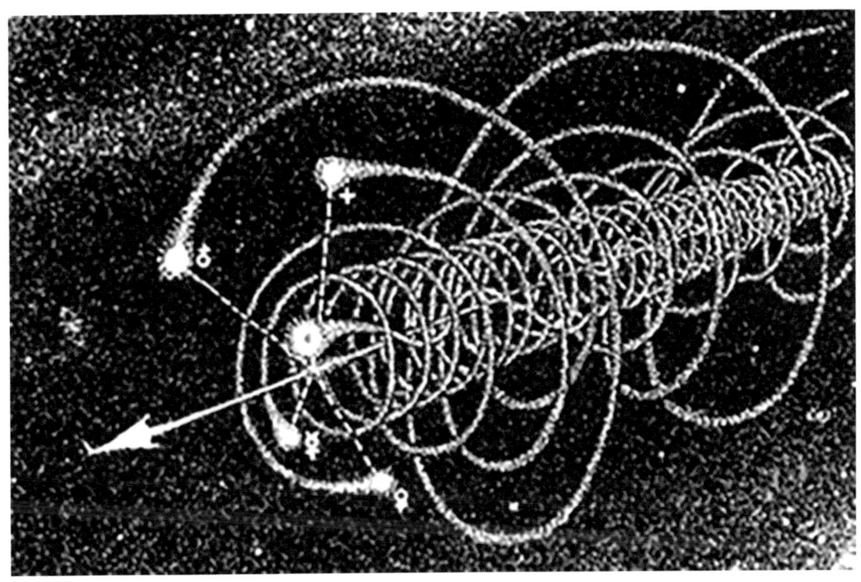

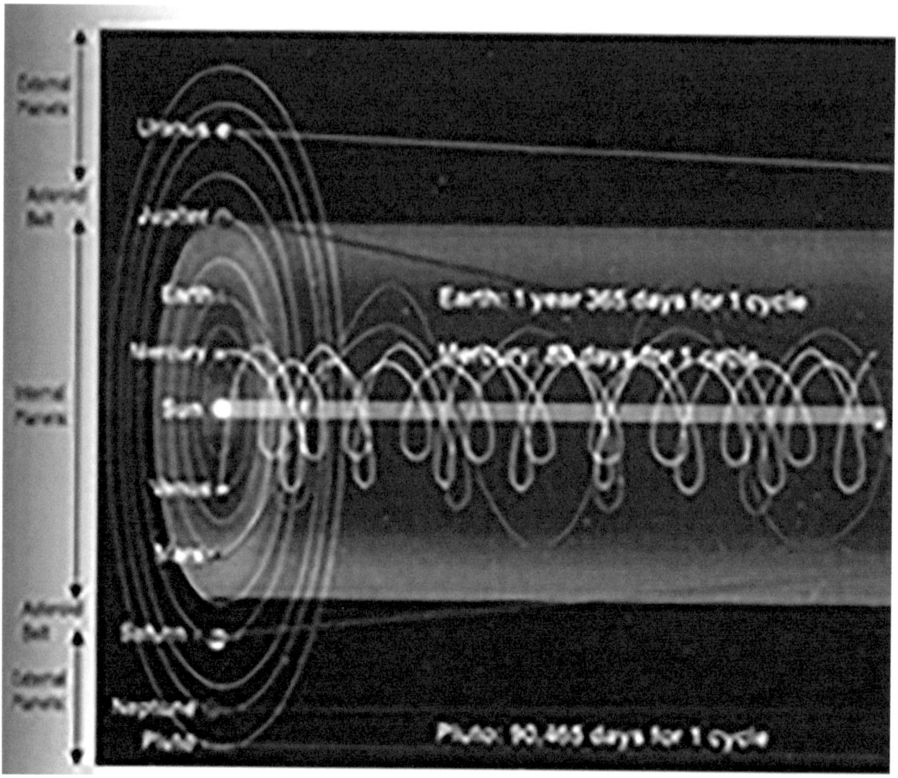
Uranus
Jupiter
Earth
Mercury
Sun
Saturn
Neptune
Pluto
Earth: 1 year 365 days for 1 cycle
Pluto: 90,465 days for 1 cycle

Planets revolve around the Sun in elliptical orbits.

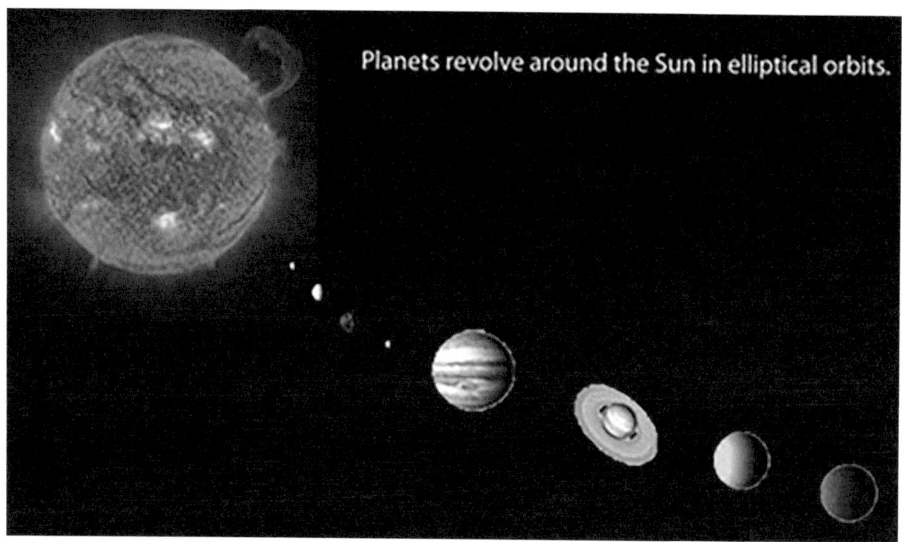

INFO you already know:

- Earth only takes twenty-three hours, fifty-six minutes and 4.09 seconds to turn once on its axis. Astronomers call this a sidereal day.
- Seen from above the north pole, the Earth rotates in a counterclockwise direction. This is why regions in the east see the sun before regions in the west.
- The Earth isn't a perfect sphere. Instead, it's an oblate spheroid—a squished ball—because the Earth's bulge points on the equator are a few kilometers more distant from the center of the Earth than the poles.
- The Earth's axis is tilted about 23.5 degrees out of vertical, and that's why we have our seasons. On the June twenty-first northern solstice, the Earth's north pole tilts most toward the sun for the year. Then, six months later, at the December twenty-second southern solstice, it's the Earth's south pole that most tilts toward the sun.
- The Earth revolves around the sun in a nearly circular orbit rather than elliptical because the minimum distance across Earth's orbit is about 98.6% of the maximum distance. That's nearly circular and not elliptical.
- The Earth revolves around the sun once in 365 days, five hours, forty-eight minutes, and forty-six seconds
- The planets move around the sun in elliptical orbits with the sun at one focus. The point in the orbit at which the planet is closest to the sun is called perihelion, and the point at which it is farthest is called aphelion.

Many of us have been taught about how the solar system works by viewing a physical model that has the sun in the middle with the planets going around and around in a simple circular orbit without properly accounting for the motion of the sun (approx. four hundred and fifty thousand miles per hour). The old model might make one picture be back where you started after a year of time has past, when in fact, you are over eleven BILLION miles from where you were a year ago! This is because the sun revolves around the galactic center.

Earth revolves around the sun, true or false?

Answer:

Believe it or not, there is no empirical evidence that Earth actually orbits the sun! The theory that Earth revolves around the sun leads to the most accurate explanation of the motions we see of all objects in the sky, and it predicts observations that are found when we look for them. Bu it's still "only a theory." If you provide an argument that logically undermines this theory, or another theory that explains what we see in the sky more accurately than this one does, then this theory will be discarded or modified to agree with yours.

This theory has been accepted for about five hundred years. It can never be "proven," but it can be disproved in a minute if you bring in the right kind of new information. That's how science works.

According to Nassim Haramein:

Earth Is Not Orbiting the Sun as We Are Taught

Because of the sun's motion, the planets make those corkscrew helical motions through the galaxy, and therefore heliocentrism is wrong.

In other words, Earth doesn't orbit the sun while the sun stays fixed in space. The Earth and the planets go around the sun, at the same time when the sun is also moving around the galaxy, and the whole cluster moves around the galaxy as a unit. That means sometimes the planets are ahead of the sun and sometimes behind it along that galactic orbit. It seems that Earth returns to the same starting point after 365 days, when in fact, you are over eleven BILLION miles from where you were a year ago!

It is also true to say that if we model the motion of Earth through space over a large number of years, it would be a helical pattern. Our sun is not the center of the galaxy. It revolves around the galactic center.

Therefore, the planets' revolution is spiral rather than circular.

The analysis of this motion shows that the planets move independently of the sun. Trajectory of motion of the planets is more complicated than the circle and not fit into the theory of revolution of something around something. The nature and trajectory of the **HELICAL REVOLUTION** indicate the existence of invisible energy, which causes the planets to revolve spirally. Partially, this energy is known to us as atomic energy. The motion of electrons is on the same principle. Planets in the solar system revolve under the action of invisible energy. This energy is revolving in a spiral, carrying away for itself material objects.

Chapter Two
Mt. Chimborazo in Ecuador Is Higher Than Mt. Everest by 2,150 Meters / 7,054 Feet

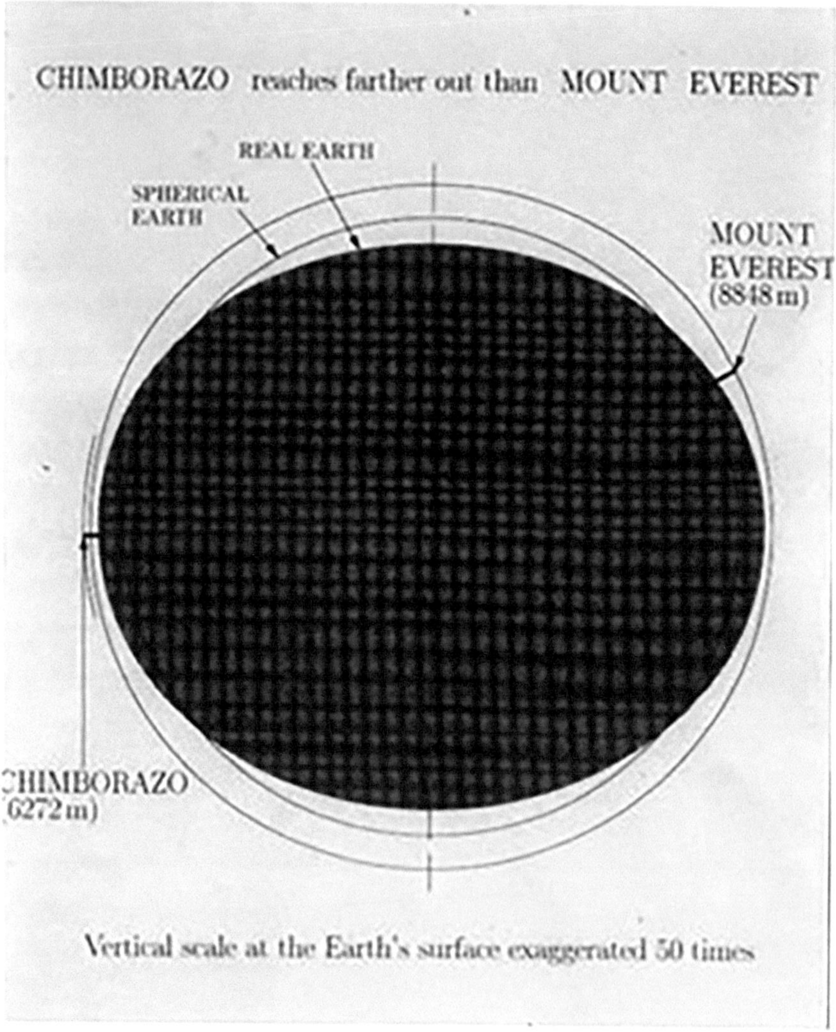

CHIMBORAZO reaches farther out than MOUNT EVEREST

REAL EARTH

SPHERICAL EARTH

MOUNT EVEREST (8848 m)

CHIMBORAZO (6272 m)

Vertical scale at the Earth's surface exaggerated 50 times

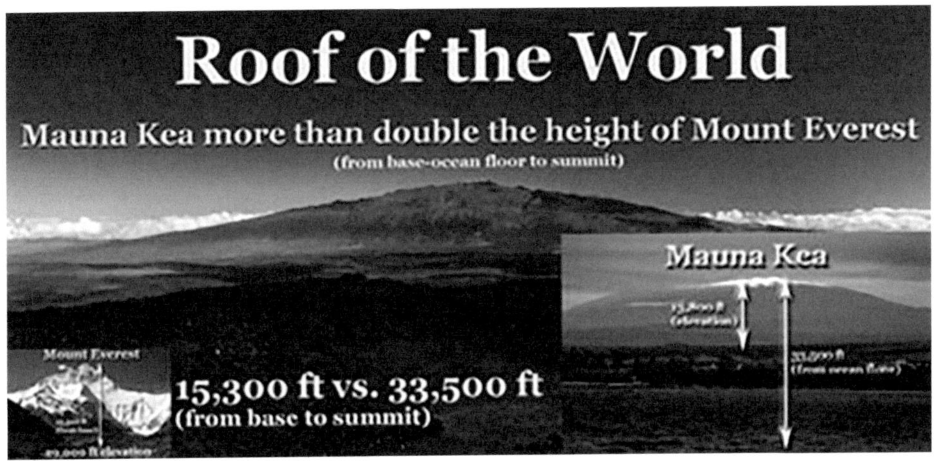

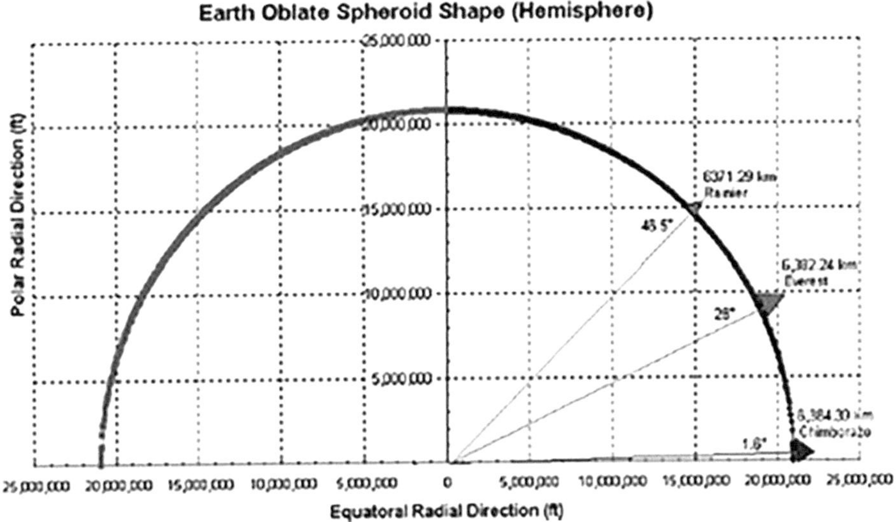

The January 2002 issue of National Geographic magazine printed an interesting confirmation of the validity of Billy Meier's claim back in the 1970s, "Mt Everest is not the highest mountain on Earth." (*Nexus New Times*, July-August 2002, page 61, www.nexusmagazine. com) Meier, in his writings, stated that Mt. Chimborazo in Ecuador was higher than Mt. Everest by 2,150 meters / 7,054 feet because the Earth is not perfectly round but, rather, bulges in the middle. Thus,

Base-to-peak heights of Mount Everest and Mount McKinley (Denali)

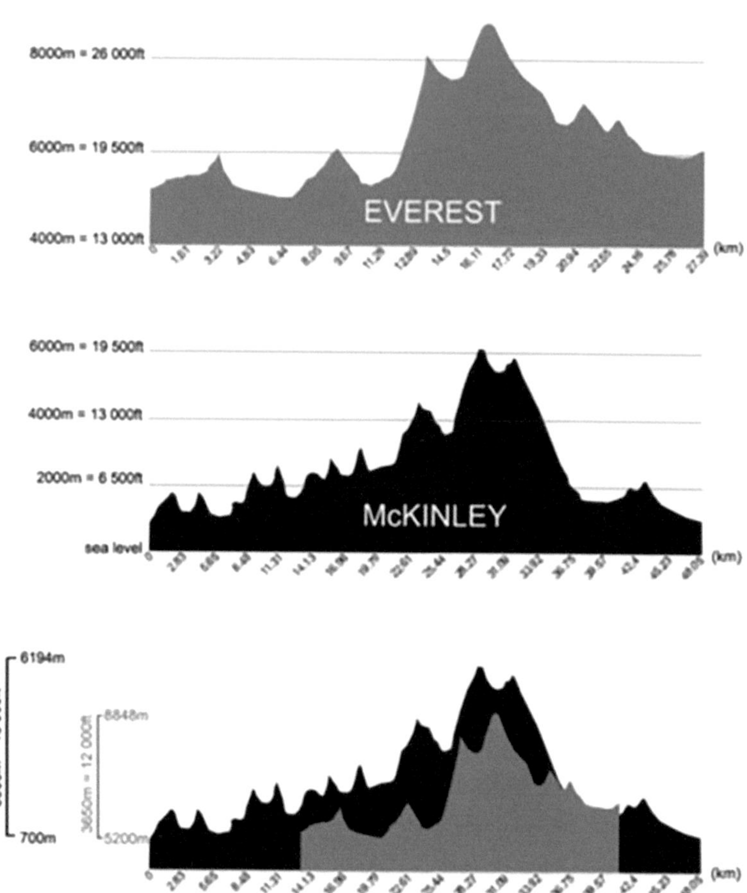

measuring mountains from sea level is not an accurate way of assessing the true height of a mountain.

National Geographic states that scientists have now determined that the Earth bulges around the middle because of the spinning action of the Earth's rotation, and, thus, when measured from the center of the planet, Mt. Chimborazo is actually higher than Mt. Everest by 2,200 meters. (*Nexus New Times*, July-August 2002, page 61, www.nexusmagazine.com) Measured from sea level, Mt. Everest is 2,540 meters higher than Mt. Chimborazo. The news brief states that when measured from the center of the Earth, Mt. Chimborazo is 6,384,450 meters high and Mt. Everest is 6,382,250 meters high.

Chapter Three
Sizes of China, India, Mexico and Africa Are Grossly Wrong on Google World Map Based on Mercator Projection

THE WORLD

WORLD

"Each model, like each map, can only show part of the truth. And world maps, make that absolutely clear. To understand "the truth" we need many models, many maps, and many points-of-view." (Dr. George F. Simons, *EuroDiversity: Managing Cultural Differences*)

<u>According to Lloyd Pye, author of "Everything You Know is Wrong":</u>

- Distortions exist to create a two-dimensional map from a three-dimensional globe.
- Google world map, according to Mercator Projection, has a grossly European centrist viewpoint.
- 18.9 million square miles of the **Northern Hemisphere** north of the equator is shown much larger than the 38.6 million square miles of the **Southern Hemisphere** south of the equator.
- 3.93 million square miles of **Europe** is shown much larger than the 6.9 million square miles of **South America**.
- 0.8 million square miles of **Greenland** is shown much larger than the 3.7 million square miles of **China**, which should be about four times the size of **Greenland**.
- 0.4 million square miles of **Scandinavia** is shown dwarfing 1.3 million square miles of **India**, which should be three times larger than **Scandinavia**.
- 0.6 million square miles of **Alaska** are drawn much larger than 0.7 square miles of **Mexico**.
- 8.6 million square miles of the former **USSR** is greatly inflated in contrast to the greatly diminished 11.6 million square miles of **Africa**.

•**<u>According to Dr. George F. Simons, author of Euro Diversity (Managing Cultural Differences):</u>**

- **Indonesia**'s total land area is 741,052 sq mi. It is hard to see from the map that it is about five times that of **Japan**'s total land area of 145,000 sq. miles. In fact, **Japan** is smaller than the **State of Montana, United States**, which is 147,000 sq. miles.

- Many of the traditional maps like the Mercator exaggerated the goodies for their primary consumers, the developed nations of the Northern Hemisphere, and gave us a Euro/ North American centric world, exactly the problem we are trying to deal with as we train our people to "go global" and come to grips with other folks in our extended human enterprises.
- The distance between many maps and the reality they aim to represent may have resulted historically from a combination of ignorance, self-interest, limited perspectives, and in some cases even deliberate deception. Looking at ancient maps, we often notice that familiar territory is highly detailed and relatively accurate while faraway places with strange sounding names were often quite lumpy and indistinct, even as more remote regions earned the name terra incognita—unknown land.
- Maps are in a sense like two-dimensional photographs, always true but at the same time always deceptive, because they represent a moment in time and can only display two dimensions at a time. And even those two dimensions are not typically on a uniform scale or common unit of measure.
- Sizes of countries and distances in our world are of course best represented by a globe, so attempts at creating flat maps inevitably yield distortions. The old Mercator was designed for navigation and was perfect for directions (lines of constant compass bearing). It also shows shapes somewhat accurately. However, the Mercator distorted the sizes of the land masses terribly.
- "What does it mean if you are not represented on the map?"

McArthur's Universal Corrective Map of the World / South Up Map

Who says north has to be on top?

Stuart McArthur of Melbourne, Australia produced the world's first "modern" south up map and launched it on Australia Day in 1979. Not only does it have south at the top, but it is also "rotated" so that China, Indonesia, and Australia are in the center rather than Europe and West Africa.

Chapter Four
Sun Worshipping Cultures Do NOT Worship Our Sun

The concept of sun worship is one nearly as old as mankind itself. We have been taught that.

- We have Sunday in honor of the sun god.
- The Egyptian peoples honored Ra, the sun god.
- The Greeks honored Helios.
- Native American tribes perform a sun dance each year.
- Ancient Persian societies celebrated the rising of the sun each day.
- Chinese and Korean used the Sun as the ultimate symbol in their philosophy, the concept of yin-yang (light and dark)

However, according to Wayne Herschel, author and symbologist who deciphers Ra Sun symbols in ancient star maps and metal book codices in Jordan: It is not our sun, Solaris, that these ancient cultures worshipped. Historians were mistaken. The ancients revered a sun.

For our ancestors the summer solstice and winter solstice were a time of great spiritual import. However, it was not our sun at all. It was another sun-like star system. Herschel also further states that the gods of the ancients originated from other star systems. And they left star maps for our ancestors, starting in Egypt.

Precession of the equinoxes is a slow westward shift of the equinoxes along the plane of the ecliptic, resulting from the precession of the Earth's axis of rotation and causing the equinoxes to occur earlier each sidereal year. The precession of the equinoxes occurs at a rate of 50.27 seconds of arc a year; a complete precession requires 25,800 years. Wayne Herschel discovered thirty-eight ancient alien star maps around the world and inferred that alien beings from other star systems taught our ancestors knowledge of mathematics, physics, and astronomy. Ancient humans simply did not have the technology and life-span to observe and document the precession of the equinoxes. It is one of the important legacies and pieces of knowledge that came from alien star beings.

Chapter Five
"X-Men" Superhumans Could Become a Reality in Thirty Years Say MoD Experts

The following is an excerpt from Daily Mail Online by James Rush. Published on 25 February 2013.

Advancements in gene technology could help humans gain mutant powers such as the likes of Wolverine, Cyclops, and Storm in the popular comic book and movie series, it has been reported.

The MoD's Development Concepts and Doctrine Centre warn however that "genetic inequality" could result from advancements in biology being unequally shared across society.

The center met last summer for a two-day summit, featuring experts from government, industry, and universities. The details have been released following a freedom of information request by *The Sun*. It was reported during the summit, held to predict what would happen in the future, "Advancements in gene technology could lead to a class of genetically superior humans by 2045."

"Human augmentation is likely to increase over the next thirty years." "Discussions highlighted that it is possible that advances in biology, unequally shared across society, could generate genetic inequality.

The X-Men are a team of mutant superheroes created by writer Stan Lee and artist Jack Kirby that first appeared in Marvel Comics in 1963. The mutants use their powers for the benefit of humanity, despite an ever-growing anti-mutant sentim, ent among mankind. The comics were turned into a highly successful film series, featuring Hugh Jackman as Wolverine, Halle Berry as Storm, Ian McKellan as Magneto and Patrick Stewart as Professor X. Halle Berry, playing the character Storm in the popular film series, is able to use her powers to manipulate the weather. Professor X, played by Patrick Stewart in the films, is known as the leader and founder of the X-Men and is able to read, control, and influence human minds.

Chapter Six
Eisenhower's Greada Treaty with the Aliens in 1954

Back in 1954, under the Eisenhower administration, the federal government decided to circumvent the Constitution of the United States and form a treaty with alien entities. The treaty does exist and there are many witnesses to the meeting between aliens and President Eisenhower. Laura Magdalene Eisenhower, the great granddaughter of Pres. Eisenhower also pointed out much circumstantial evidence. Other evidence has been obtained, which is kept in the Library of Congress, proving the meeting with aliens happened, but documentation of the treaty signed by Ike is still many levels above top secret.

This treaty stated the aliens would not interfere in our affairs and we would not interfere in theirs. We would keep their presence on Earth secret; they would furnish us with advanced technology. They could abduct humans on a limited basis for the purpose of medical examination and monitoring, with the stipulation that the humans would not be harmed, would be returned to their point of abduction, and that the humans have no memory of the event. It was also agreed the alien bases would be constructed underground, beneath Indian reservations in the four corners area of Utah, New Mexico, Arizona, and Colorado. Another was to be constructed in Nevada in the area known as S-4, about seven miles south of Area 51, known as "Dreamland." (Michael E. Salla. "False Flag Event in Syria: The Third force and Exopolitics." Posted in "Archaeology, Exopolitics Activism, World Politics," September 9, 2013. http://exopolitics.org/author/dr- michael-salla/)

Chapter Seven
Nature's Favorite Numbers—Fibonacci Numbers

The Fibonacci Sequence is the series of numbers:

0, 1, 1, 2, 3, 5, 8, 13, 21, 34 . . .

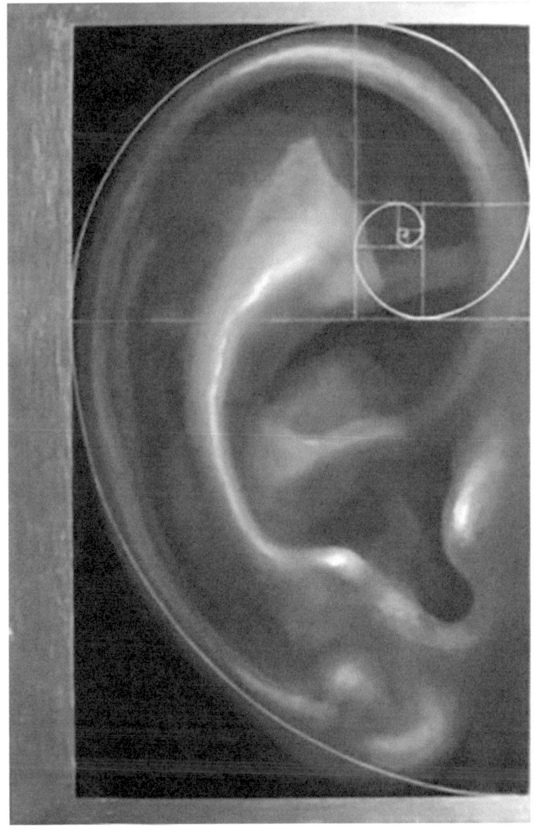

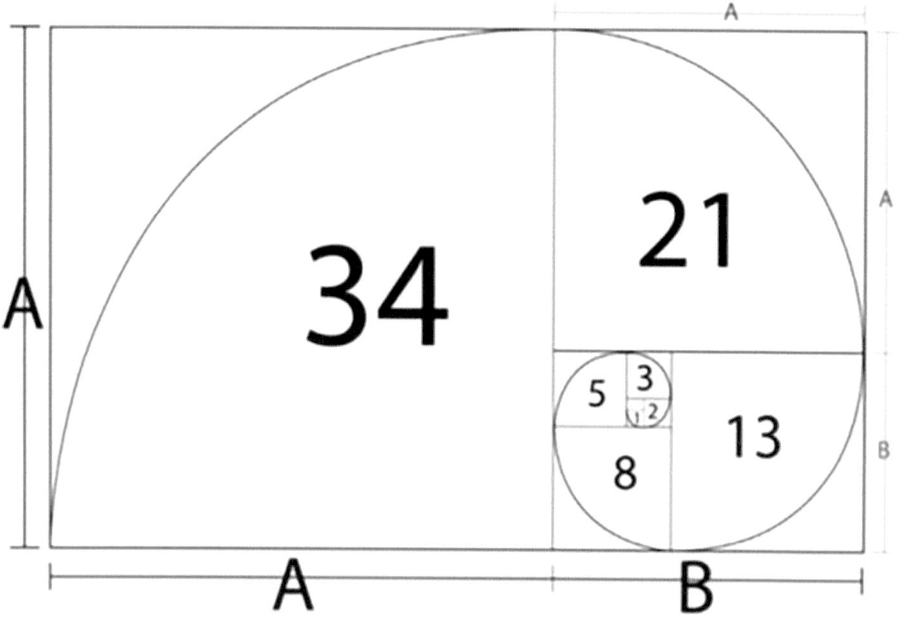
A
A
34
21
A
5
3
2
1
8
13
B
A
B

The next number is found by adding up the two numbers before it. The two is found by adding the two numbers before it (1+1)

Similarly, the three is found by adding the two numbers before it (1+2), And the five is (2+3), and so on!

The Fibonacci numbers are Nature's numbering system. They appear everywhere in Nature, from the leaf arrangement in plants, to the pattern of the florets of a flower, the bracts of a pinecone, or the scales of a pineapple. The Fibonacci numbers are therefore applicable to the growth of every living thing, including a single cell, a grain of wheat, a hive of bees, and even all of mankind.

Plants do not know about this sequence - they just grow in the most efficient ways. Many plants show the Fibonacci numbers in the arrangement of the leaves around the stem. Some pinecones and fir cones also show the numbers, as do daisies and sunflowers. Sunflowers can contain the number 89, or even 144. Many other plants, such as succulents, also show the numbers. Some coniferous trees show these numbers in the bumps on their trunks. And palm trees show the numbers in the rings on their trunks.

Why do these arrangements occur? In the case of leaf arrangement, or phyllotaxis, some of the cases may be related to maximizing the space for each leaf, or the average amount of light falling on each one. Even a tiny advantage would come to dominate, over many generations. In the case of close-packed leaves in cabbages and succulents the correct arrangement may be crucial for availability of space.

In the seeming randomness of the natural world, we can find many instances of mathematical order involving the Fibonacci numbers themselves and the closely related "Golden" elements.

Fibonacci Petals

3 petals: lily, iris

5 petals: buttercup, wild rose, larkspur, columbine

8 petals: delphiniums

13 petals: ragwort, corn marigold, cineraria

21 petals: aster, black-eyed susan, chicory

34 petals: plantain, pytethrum

55, 89 petals: michelmas daisies, the asteraceae family

The occurrence of Fibonacci Numbers in Nature is interesting but the ratio of consecutive Fibonacci Numbers is important.

Fibonacci in Animals

The shell of the Chambered Nautilus has golden proportions. It is a logarithmic spiral.

The eyes, fins, and tail of the dolphin fall at Golden sections along the body.

A starfish has five arms. (Five is the fifth Fibonacci number) If a regular pentagon is drawn and diagonals are added, a five-sided star or pentagram is formed. Where the sides of the pentagon are one unit in length, the ratio between the diagonals and the sides is Phi, or the Golden Ratio. This five-point symmetry with Golden proportions is found in starfish.

Humans exhibit Fibonacci characteristics, too. The golden ratio is seen in the proportions of the sections of a finger.

It is also worthwhile to mention that we have 8 fingers in total, five digits on each hand, three bones in each finger, two bones in one thumb, and one thumb on each hand.

The ratio between the forearm and the hand is the Golden Ratio.

The cochlea of the inner ear forms a Golden Spiral.

(Nikhat Parveen, “Fibonacci in Nature,” University of Georgia. http://jwilson.coe.uga.edu/emat6680/parveen/fib_nature.htm)

Chapter Eight
Sonic Levitation, Megaliths, and Pyramids, City of Jericho Destroyed by Sonic Vibrations

"In almost every culture where megaliths exist, according to page 432: *Cosmic Key*, a legend also exists that the huge stones were moved by acoustic means—either by the chanted spells of magicians, by song, by striking with a magic wand or rod (to produce acoustic resonance), or by trumpets, gongs, lyres, cymbals or whistles." (Stephen Wagner. *The Ancient Secrets of Levitation.* About.com. http://paranormal.about.com/od/antigravity/a/The-Ancient-Secrets-Of-Levitation.htm)

Did ancient civilizations possess knowledge that has since been lost to science? Were amazing technologies available to the ancient Egyptians that enabled them to construct the pyramids—technologies that have somehow been forgotten?

The ruins of several ancient civilizations—from Stonehenge to the pyramids—show that they used massive stones to construct their monuments. A basic question is why? Why use stone pieces of such enormous size and weight when the same structures could have been constructed with more easily managed smaller blocks, much like we use bricks and cinderblocks today?

Could part of the answer be that these ancients had a method of lifting and moving these massive stones—some weighing several tons—that made the task as easy and manageable as lifting a two-pound brick?

The ancients, some researchers suggest, may have mastered the art of levitation, through sonics or some other obscure method, which allowed them to defy gravity and manipulate massive objects with ease.

Acoustic levitation

"Acoustic levitation uses sound traveling through a fluid—usually a gas—to balance the force of gravity. On Earth, this can cause objects and materials to hover unsupported in the air. In space, it can hold objects steady so they don't move or drift. The process relies on of the properties of sound waves, especially intense sound waves." (Tracy V. Wilson. *How Acoustic Levitation Works*. http://science.howstuffworks. com/acoustic-levitation1.htm)

The Physics of Sound Levitation

A basic acoustic levitator has two main parts: a transducer, which is a vibrating surface that makes sound, and a reflector. Often, the transducer and reflector have concave surfaces to help focus the sound. A sound wave travels away from the transducer and bounces off the reflector. Three basic properties of this traveling, reflecting wave help it to **suspend objects in midair.**

- First, the wave, like all sound, is a longitudinal pressure wave. In a longitudinal wave, movement of the points in the wave is parallel to the direction the wave travels. It's the kind of motion you'd see if you pushed and pulled one end of a stretched Slinky. Most illustrations, though, depict sound as a transverse wave, which is what you would see if you rapidly moved one end of the Slinky up and down. This is simply because transverse waves are easier to visualize than longitudinal waves.

- Second, the wave can bounce off surfaces. It follows the law of reflection, which states that the angle of incidence—the angle at which something strikes a surface—equals the angle of reflection—the angle at which it leaves the surface. In other words, a sound wave bounces off a surface at the same angle at which it hits the surface. A sound wave that hits a surface head on at a ninety-degree angle will reflect straight back off at the same angle. The easiest way to understand wave reflection is to imagine a Slinky that is attached to a surface at one end. If you picked up the free end of the Slinky and moved it rapidly up and then down, a wave would travel the length of the spring. Once it reached the fixed end of the spring, it would reflect off the surface and travel back toward you. The same thing happens if you push and pull one end of the spring, creating a longitudinal wave.

• Finally, when a sound wave reflects off a surface, the interaction between its compressions and rarefactions causes interference. Compressions that meet other compressions amplify one another, and compressions that meet rarefactions balance one another out. Sometimes, the reflection and interference can combine to create a standing wave. Standing waves appear to shift back and forth or vibrate in segments rather than travel from place to place. This illusion of stillness is what gives standing waves their name.

Standing sound waves have defined nodes, or areas of minimum pressure, and antinodes, or areas of maximum pressure. A standing wave's nodes are at the heart of acoustic levitation. Imagine a river with rocks and rapids. The water is calm in some parts of the river, and it is turbulent in others. Floating debris and foam collect in calm portions of the river. In order for a floating object to stay still in a fast-moving part of the river, it would need to be anchored or propelled against the flow of the water. This is essentially what an acoustic levitator does, using sound moving through a gas in place of water.

By placing a reflector, the right distance away from a transducer, the acoustic levitator creates a standing wave. When the orientation of the wave is parallel to the pull of gravity, portions of the standing wave have a constant downward pressure and others have a constant upward pressure. The nodes have very little pressure.

In space, where there is little gravity, floating particles collect in the standing wave's nodes, which are calm and still. On Earth, objects collect just below the nodes, where the acoustic radiation pressure, or the amount of pressure that a sound wave can exert on a surface, balances the pull of gravity.

Objects hover in a slightly different area within the sound field depending on the influence of gravity. It takes more than just ordinary sound waves to supply this amount of pressure.

Ed Leedskalnin and Coral Castle

Ed Leedskalnin said science is misled by this false electron/proton principle. He said there are no electrons as perceived. To quote Ed: "The trouble with physicists is they use indirect and ultra indirect methods to come to their conclusions."

Coral Castle is a stone structure created by the Latvian American eccentric Edward Leedskalnin (1887–1951) north of the city of Homestead, Florida in Miami-Dade County. The structure comprises numerous megalithic stones (mostly limestone formed from coral), each weighing several tons. It currently serves as a privately operated tourist attraction. Coral Castle is noted for legends surrounding its creation that claim it was built single-handedly by Leedskalnin using reverse magnetism and/or supernatural abilities to move and carve numerous stones weighing many tons.

Two magnetic currents are the fabric of what is holding together everything in our universe, including the atoms themselves. This is what builds and holds an atom into a structure. The ancients knew this fantastic secret, as did Ed Leedskalnin, and there is a ton of evidence. The message here not to be ignored is Ed Leedskalnin had amazing insight and has given us the true meaning of sacred geometry, and that it is a representation of two magnetic currents. And he left us the visual message before he died.

Biblical city of Jericho B.C. 1400

"An acoustic weapon disorients rioters and afflicts an invading army with nausea. It can create 'ghosts' and arouse animal passions.

"Possibly the earliest account in Western literature of sound itself being used as a weapon can be found in the Bible. As detailed in Joshua 6:5, Joshua leads an attack on the city of Jericho during which he commands his people, outside the walled city, to remain in total silence for seven days. On the seventh day, seven trumpets made from ram's horns give a "long blast," the people shout . . . and the walls of Jericho come crashing down.

"Sound is a waveform, with low infrasonic frequencies having a long wave length (measured in tens of meters), and with high ultrasonic frequencies having a short wave length (measured in millimeters). The frequencies associated with ultrasound are most familiar from their utilization by the medical profession, chiefly for diagnostic imaging.

"While the ears are designed to detect a limited range of frequencies – the human auditory range is between 20Hz and 20,000Hz (1Hz =1 cycle per second) – different frequencies can affect the whole body and, at volume, can be felt in almost any part of the body. Even with industrial ear protectors, sound waves are able to enter the head via the nose and mouth that are, in turn, linked to the ears by the structure of the skull. Sounds that are higher in frequency than 20,000Hz – ultrasound – are inaudible to humans, while sounds lower than 20Hz– infrasound – are inaudible but can, on occasion, be felt resonating within the body itself. Exposure of unprotected ears to infrasound can also cause an increase in pressure within the middle ear, disturbing the sense of balance." (Jack Sargeant. "Sonic Weapons." Forteantimes. com. December 2001. http://www.forteantimes.com/features/ articles/256/sonic_weapons.html)

Chapter Nine
The Electric Universe Theory

Today, nothing is more important to the future and credibility of science than liberation from the gravity-driven universe of prior theory. A mistaken supposition has not only prevented intelligent and sincere investigators from seeing what would otherwise be obvious, it has bred indifference to possibilities that could have inspired the sciences for decades.

~*David Talbott and Wallace Thornhill,* Thunderbolts of the Gods

The Electric Universe

Posted on August 23, 2011 by David Talbott

Readers may be surprised to discover that many well-trained skeptics do not support popular ideas in astronomy and the space sciences. Critics doubt that "black holes" actually exist. They suggest that "dark matter," supposedly far more abundant than visible matter, is a mere fiction, hiding the fact that earlier theories no longer work. Theories of galaxy formation, the birth of stars, and the evolution of our planetary system are all raised to doubt by critics who believe that a fateful turn in twentieth century theory set astronomy on a dead-end course.

Enchanted by the role of gravity in the cosmos, astronomers failed to recognize the pervasive role of charged particles and electric currents in space. Electric Universe is a new vantage point, one that acknowledges the contribution of the electric force to the dynamic structure and highest energy events in the universe. As we compare events in space to the behavior of charged particles in the laboratory, the differences between an electric model and the traditional gravity only model should become progressively more clear.

Essential Guide to the Electric Universe

Posted on September 2, 2011 by Bob Johnson - Jim Johnson

The New Picture of Space

Now more than ever, the exploration of our starry universe excites the imagination. Never before has space presented so many pathways for research and discovery.

New observational tools enable us to "see" formerly invisible portions of the electromagnetic spectrum, and the view is spectacular. Telescope images in X-ray, radio, infrared and ultraviolet light reveal exotic structure and intensely energetic events that continually redefine the quest as a whole.

Spectrographic interpretation has grown hand in hand with faster, large memory computers and programs in sophistication and in broad scientific data processing, imaging, and modeling capability.

Standing out amidst an avalanche of new images is the greatest surprise of the space age: evidence for pervasive electric currents and magnetic fields across the universe, all connecting and animating what once appeared as isolated islands in space. The intricate details revealed are not random, but exhibit the unique behavior of charged particles in plasma under the influence of electric currents.

The telltale result is a complex of magnetic fields and associated electromagnetic radiation. We see the effects on and above the surface of the sun, in the solar wind, in plasma structures around planets and moons, in the exquisite structure of nebulas, in the high energy jets of galaxies, and across the unfathomable distances between galaxies.

Thanks to the technology of the twentieth century, astronomers of the twenty-first century will confront an extraordinary possibility. The evidence suggests that intergalactic currents, originating far beyond the boundaries of galaxies themselves, directly affect galactic evolution. The observed fine filaments and electromagnetic radiation in intergalactic and interstellar plasma are the signature of electric currents. Even the power lighting the galaxies' constituent stars may indeed be found in electric currents winding through galactic space.

It was long thought that only gravity could do "work" or act effectively across cosmic distances. But perspectives in astronomy are rapidly changing. Specialists trained in the physics of electricity and magnetism have developed new insights into the forces active in the cosmos. A plausible conclusion emerges. Not gravity alone, but electricity and gravity have shaped and continue to shape the universe we now observe.

The Limits of Gravitational Theory

The Law of Gravity, which relies exclusively on the masses of celestial bodies and the distances between them, works very well for explaining planetary and satellite motions within our solar system. But when astronomers tried to apply it to galaxies and clusters of galaxies, it turns out that nearly 90 percent of the mass necessary to account for the observed motion is missing.

The trouble began in 1933 when astronomer Fritz Zwicky calculated the mass-to-light ratio for eight galaxies in the Coma Cluster of the Coma Berenices ("Berenices's hair") constellation. At the time, it was assumed that the amount of visible light coming from stars should be proportional to their masses (a concept called "visual equilibrium").

As Zwicky was to realize, the apparent rapid velocities of the galaxies, around their common center of mass ("barycenter"), suggested that much more mass than could be seen was required to keep the galaxies from flying out of the cluster.

Zwicky concluded that the missing mass must therefore be invisible or "dark." Other astronomers, such as Sinclair Smith (who performed calculations on the Virgo Cluster in 1936) began to find similar problems. To make matters worse, in the 1970s, radial velocity plots (radius from the center versus stars' speed of rotation) for stars in the Milky Way galaxy revealed that the speeds flatten out rather than trail down, implying that velocity continues to increase with radius, contrary to what Newton's Law of Gravity predicts for, and which is observed in, the Solar System.

In short, astronomers using the gravity model were forced to add a lot more mass to every galaxy than can be detected at any wavelength. They called this extra matter "dark;" its existence can only be inferred from the failure of predictions. To cover for the insufficiency, they gave themselves a blank check, a license to place this imagined stuff wherever needed to make the gravitational model work.

Other mathematical conjectures followed. Assumptions about the red shift of objects in space led to the conclusion that the universe is expanding. Then other speculations led to the notion that the expansion is accelerating. Faced with an untenable situation, astronomers postulated a completely new kind of matter, an invisible "something" that repels rather than attracts. Since Einstein equated mass with energy ($E = mc^2$), this new kind of matter was interpreted as being of a form of mass that acts like pure energy—regardless of the fact that if the matter has no mass it can have no energy according to the equation. Astronomers called it "dark energy," assigning to it an ability to overcome the very gravity on which the entire theoretical edifice rested.

Dark energy is thought to be something like an electrical field, with one difference. Electric fields are detectable in two ways: when they accelerate electrons, which emit observable photons as synchrotron and Bremsstrahlung radiation, and by accelerating charged particles as electric currents, which are accompanied by magnetic fields, detected through Faraday rotation of polarized light. Dark energy seems to emit nothing and nothing it purportedly does is revealed through a magnetic field. One suggestion is that some property of empty space is responsible. But empty space, by definition, contains no matter and therefore has no energy. The concept of dark energy is philosophically unsound and is a poignant reminder that the gravity only model never came close to the original expectations for it.

Taking the postulated dark matter and dark energy together, something on the order of twenty-four times as much mass in the form of invisible stuff would have to be added to the visible, detectable mass of the universe. That's to say, in the gravity model all the stars and all the galaxies and all the matter between the stars that we can detect only amount to a minuscule 4 percent of the estimated mass:

Critics often point out that a theory requiring speculative, undetectable stuff on such a scale also stretches credulity to the breaking point. Something very real, perhaps even obvious, is almost certainly missing in the standard gravity model.

Is it possible that the missing component could be something as familiar to the modern world as electricity?

Chapter Ten
Directed Panspermia Proposed by Francis Crick

To overcome the huge hurdles of evolution of life from non-living chemicals on Earth, Crick proposed, in a book called *Life Itself*, that some form of primordial life was shipped to the Earth billions of years ago in spaceships—by supposedly "more evolved" alien beings.

Michael Denton says in his best-selling book *Evolution a Theory in Crisis*, "Nothing illustrates clearly just how intractable a problem the origin of life has become than the fact that world authorities can seriously toy with the idea of Panspermia."

The following is taken from Panspermia-Theory.com, "Recent Support for Panspermia." http://www.panspermia-theory.com/

A meteorite blasted off from the surface Mars about 15 million years ago was found in Antarctica in 1984 by a team of scientists on an annual United States government mission to search for meteors. The meteor was named Allan Hills 84001 (ALH84001). In 1996 ALH84001 was shown to contain structures that may be the remains of terrestrial nanobacteria. The announcement, published in the journal Science by David McKay of NASA, made headlines worldwide and prompted United States President Bill Clinton to make a formal televised announcement marking the event and expressing his commitment to the aggressive plan in place at the time for robotic exploration of Mars. Several tests for organic material have been performed on ALH84001 and amino acids and polycyclic aromatic hydrocarbons (PAH) have been found. However, most experts now agree that these are not a definite indication of life, but may have instead been formed abiotically from organic molecules or are due to contamination from contact with Antarctic ice. The debate is still ongoing, but recent advances in nanobe research has made the find interesting again.

The announcement of the discovery of evidence of life on ALH84001 sparked a surge in support for the theory of Panspermia. People began to speculate about the possibility that life originated on Mars and was transported to Earth on debris ejected after major impacts. On April 29, 2001, at the 46th annual meeting of the International Society for Optical Engineering (SPIE) in San Diego, California, Indian and British researchers headed by Chandra Wickramasinghe presented evidence that the Indian Space Research Organisation had gathered air samples from the stratosphere that contained clumps of living cells. Wickramasinghe called this "unambiguous evidence for the presence of clumps of living cells in air samples from as high as 41 kilometers, well above the local tropopause, above which no air from lower down would normally be transported." A reaction report from NASA Ames doubted that living cells could be found at such high altitudes, but noted that some microbes can remain dormant for millions of years, possibly long enough for an interplanetary voyage within a solar system.

On May 11, 2001, Geologist Bruno d'Argenio and molecular biologist Giuseppe Geraci from the University of Naples announced the finding of extraterrestrial bacteria inside a meteorite estimated to be over 4.5 billion years old. The researchers claimed that the bacteria, wedged inside the crystal structure of minerals, had been resurrected in a culture medium. They asserted that the bacteria had DNA unlike any on Earth and had survived when the meteorite sample was sterilized at high temperature and washed with alcohol. The bacteria were determined to be related to modern day Bacillus subtilis and Bacillus pumilus bacteria, but appear to be a different strain.

On April 21, 2008, renowned British astrophysicist Stephen Hawking spoke about Panspermia during his "Why We Should Go Into Space" lecture for NASA's fiftieth Anniversary lecture series at George Washington University.

In a virtual presentation on Tuesday, April 7, 2009, Stephen Hawking discussed the possibility of building a human base on another planet and gave reasons why alien life might not be contacting the human race, during his conclusion of the Origins Symposium at Arizona State University. Hawking also talked about what humans may find when venturing into space, such as the possibility of alien life through the theory of Panspermia, which says that life in the form of DNA particles can be transmitted through space to habitable places.

Part 4
Be Skeptical Even of One's Own Skepticism

We all know the famous story surrounding Sir Isaac Newton. He was busy discovering the universal law of gravitation, when an apple fell on his head from the branches above him. The fallen apple provided him the inspiration for the laws of gravity. But while the falling apple is a good story, it probably never happened. The story was first published in an essay by Voltaire, long after Newton's death. It was almost certainly an invention.

Centuries-old manuscripts also revealed the hidden pursuits of Sir Isaac Newton when the program "Newton's Dark Secrets" was aired November 15, 2005 at 9 p.m. on PBS. While Isaac Newton was busy discovering the universal law of gravitation, he was also searching out hidden meanings in the Bible and pursuing the covert art of alchemy.

Discoveries were made that showed the unknown side of Newton to the rest of the world. Discoveries that showed Isaac Newton was obsessed with religion and devoted to the occult. Our modern interpretation of Newton is about as far fetch as could possibly be from what Newton himself thought.

Known as the Greatest Scientist of all times who invented calculus; figured out the composition of light; and gave us the laws of gravity and motion, which govern the universe; Sir Isaac Newton was actually pursuing an activity which we now label as a pseudoscience.

We all know the famous story surrounding Sir Isaac Newton. He was busy discovering the universal law of gravitation, when an apple fell on his head from the branches above him. The fallen apple provided him the inspiration for the laws of gravity. But while the falling apple is a good story, it probably never happened. The story was first published in an essay by Voltaire, long after Newton's death. It was almost certainly an invention.

Centuries-old manuscripts also revealed the hidden pursuits of Sir Isaac Newton when the program "Newton's Dark Secrets" was aired November 15, 2005 at 9 p.m. on PBS. While Isaac Newton was busy discovering the universal law of gravitation, he was also searching out hidden meanings in the Bible and pursuing the covert art of alchemy.

Discoveries were made that showed the unknown side of Newton to the rest of the world. Discoveries that showed Isaac Newton was obsessed with religion and devoted to the occult. Our modern interpretation of Newton is about as far fetch as could possibly be from what Newton himself thought.

Known as the Greatest Scientist of all times who invented calculus; figured out the composition of light; and gave us the laws of gravity and motion, which govern the universe; Sir Isaac Newton was actually pursuing an activity which we now label as a pseudoscience.

The wars for truths are fought relentlessly and ruthlessly over generations. Fear cannot save us. Rage cannot help us. What misfortune or bliss does ignorance hold in store for us in the next hundred years?

There is no sound, no voice, no cry in all the world that can be heard about the truth . . . until someone listens.

Chapter One
Map of Antarctica Existed, Three Years Before Its Discovery

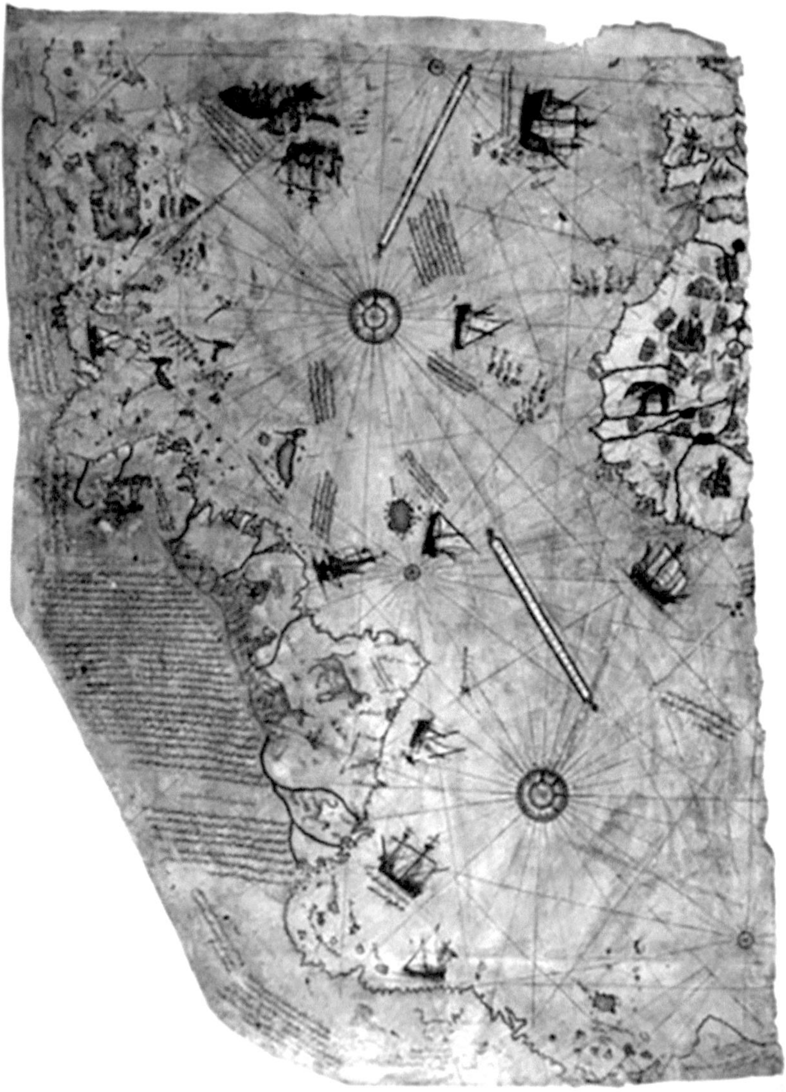

The Piri Reis map shows the western coast of Africa, the eastern coast of South America, and the northern coast of Antarctica. The northern coastline of Antarctica is perfectly detailed. The most puzzling however is not so much how Piri Reis managed to draw such an accurate map of the Antarctic region three hundred years before it was discovered, but that the map shows the coastline under the ice. Geological evidence confirms that the latest date Queen Maud Land could have been charted in an ice-free state is 4000 BC.

The question is: Who mapped the Queen Maud Land of Antarctic six thousand years ago? Which unknown civilization had the technology or the need to do that? It is well-known that the first civilization, developed in the mid-east around year 3000 BC, was soon followed within a thousand years by the Indus valley and the Chinese ones. So, accordingly, none of the known civilizations could have done such a job. Who was here in 4000 BC, able to do things that NOW are possible with modern technologies?

Chapter Two
Pre-Colombian Space Shuttle Models and Ancient Indian Flying Crafts

Pre-Columbian funerary pendants –Gold trinkets of pre-Colombian space shuttle models were found in an area covering Central America and coastal areas of South America, estimated to belong to a period between 500 and 800 CE. However, since they are made from gold, accurate dating is impossible and based essentially on stratigraphy, which may be deceptive. However, we can safely say that these gold trinkets of pre-Colombian space shuttle models are more than one thousand years old.

Clearly, these fascinating objects display aerodynamic features, complete with delta wings, tail stabilizers and elevators, fuselage, and a rudder. They are definitely not stylized figures of fish, birds, frogs, bats, or insects.

In the Vedic literature of India, there are many descriptions of flying machines that are generally called vimanas. Many of the documents contain texts that seem to describe modern aerodynamic principles. Vymaanika-Shaastra talked about metals, electricity, and power sources used for these air crafts. For vimanas pilots, they had special uniforms, weapons that were kept hidden on these air ships, and special flight manuals for references. These fall into two categories:

- Man-made craft that resemble airplanes and fly with the aid of birdlike wings. They are described mainly in medieval, secular Sanskrit works dealing with architecture, automata, military siege engines, and other mechanical contrivances.

- Unstreamlined structures that fly in a mysterious manner and are generally not made by human beings. They are described in ancient works such as the Rg Veda, the Maha- bha-rata, the Rama-yana, and the Pura-nas, and they have many features reminiscent of modern flying machines.

Like many similar literatures and artifacts found in museums, they portrayed the might of ancient aeronautic powers. As more and more ancient artifacts and petroglyphs with pictures of UFOs, wormholes, stargates and alien beings are found throughout the world, the full picture of the jigsaw puzzle soon emerges.

Chapter Three
Sunken Cities of the Caribbean, France, India, and Japan

All around the world, whether it's the Caribbean, France, India, or Japan, we find evidence of a manmade structure that lie under water. One thing to keep in mind is that most of these cities did not sink; they were submerged when the sea level rose. It's really sad that these so called hardheaded "scholars" have their heads in the sand and try to discredit these people who bring to light all of these awesome discoveries. If left to these so called "scholars" we would never have found King Tut's Tomb.

It is not the religious community that fears the undiscovered discoveries that suddenly become discovered. It is the scientific community. It is the scientists who fear being humiliated when found wrong. We know of no Stone Age civilization capable of creating such gigantic ceremonial complexes. If according to the mainstream scholars, the ancient Indian,

French, and Japanese could not have created such a structure, then who did? And how did it get there?

Could the ancient underwater cities have been built to serve as bases for ancient aliens? Could these hold the answers that mankind have been seeking for centuries? Evidence that proves we are not alone. These underwater structures and cities seem to just stand out as out of place, in time, and there are questions about them that might throw a new light on the genesis of human history.

I think it's high time the truly knowledgeable archeologists and oceanographers are given a free hand to explore and bring forth as much knowledge of these ancient treasures to be shared with mankind.

Chapter Four
Lessons for Japan and Russia from Ancient Nuclear Reactors

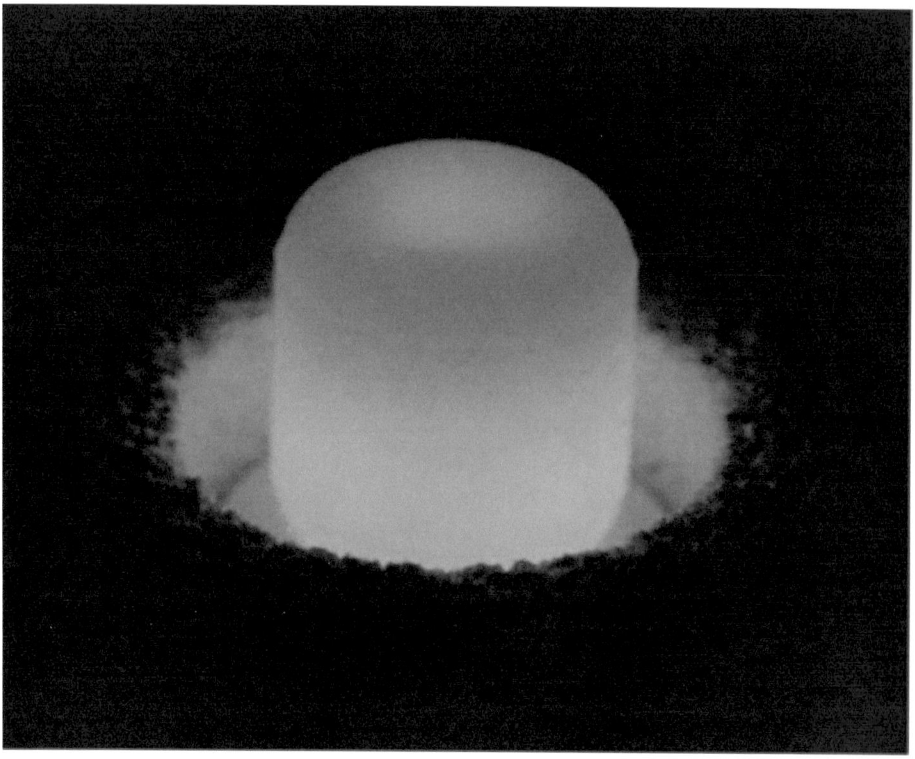

The remnants of nuclear reactors nearly two billion years old were found in 1972 in Oklo in the Gabon Republic in Africa. Surprisingly the uranium concentration in the ore was as low as spent uranium fuel from a nuclear reactor. The finding led scientists to believe that the uranium had already been used for energy production.

This discovery shocked the world and attracted scientists from many countries to go to Oklo for further investigation. The results showed that the uranium mine was an ancient nuclear reactor. The reactor was perfectly preserved, and its layout was very rational. It is estimated that the reactor had been in operation for around five hundred thousand years. Furthermore, nuclear wastes produced in this reactor had not spread all over the surrounding areas. Instead, they were confined within the separate sections. From the perspective of modern nuclear technology, this ancient reactor used very advanced techniques. Oklo by-products are being used today to study the stability of the fundamental constants over cosmological time scales and to develop more effective means for disposing of human manufactured nuclear waste.

Human beings have only made use of nuclear power for a couple of decades. This discovery raised the intriguing possibility that a technologically more advanced civilization existed two billion years ago, and it had advanced knowledge of nuclear fission.

If we neglect relics of prehistoric civilization, there is no way we can broaden the scope of our present knowledge. We will neither know what caused prehistoric civilizations to degenerate, or how they finally came to disappear. Moreover, we should carefully examine whether our current method of scientific development is following the same disastrous road. This is surely a subject worthy of serious consideration.

The long-term preservation of the Gabon natural nuclear reactors is perhaps even more remarkable than the reactors themselves. These nuclear reactors have survived two billion years of geologic time. It is certainly the most valuable lesson from ancient history if you just consider those man-made nuclear disasters like the Fukushima Daiichi nuclear disaster of 2011 and the Chernobyl disaster of 1986.

Ancient Nuclear Explosion

According to British researcher David Davenport, author of *Atomic Destruction in 2000 B.C.*, Milan, Italy, 1979:

An ancient, heavily populated city called Mohenjo Daro in Pakistan was instantly destroyed in 2,000 B.C. by an incredible explosion that could only have been caused by an atomic bomb.

The horrible, mysterious event was recorded in an old Hindu manuscript called the Mahabharata,

- White hot smoke that was a thousand times brighter than the sun rose in infinite brilliance and reduced the city to ashes.

- Water boiled . . . horses and war chariots were burned by the thousands . . .

- The corpses of the fallen were mutilated by the terrible heat so that they no longer looked like human beings . . .

- "It was a terrible sight to see . . . never before have we seen such a ghastly weapon."

What was found at the site of Mohenjo Daro corresponds exactly to Nagasaki.

- Forty-four human skeletons were found there in 1927, just a few years after the city was discovered.
- The city ruins reveal the explosion's epicenter, which measures fifty yards wide.
- At that location everything was crystallized, fused, or melted.
- "Sixty yards from the center the bricks were melted on one side indicating a blast . . . "

Davenport's intriguing theory has met with intense interest in the scientific community.

- How did man in 2000 B.C. have the knowledge of not only producing such a high degree of heat, but also harnessing the power of such high temperatures?
- If Mohanjo Daro was destroyed by a nuclear catastrophe, who designed and manufactured them?
- If not then what was used to produce such heat that vitrified rocks and bricks?
- What could be attributed to the high degree of radioactive traces in the skeletons?
- How did all of them die, in one instant?

Chapter Five

A. 7-Year-Old Russian Boy Educated People on Astronomy, Mars Landscape and How to Create High Tech Spacecraft

Sometimes, some children are born with quite fascinating talents, unusual abilities.

http://english.pravda.ru/science/mysteries/05-03-2008/104375-boriska_boy_mars-0/

A boy named Boris Kipriyanovich, or Boriska, lives in the town of Zhirinovsk of Russia's Volgograd region. He was born on January 11, 1996. Since he was four he used to visit a well-known anomalous zone, commonly referred to as Medvedetskaya Gryada – a mountain near the town. It seems that the boy needed to visit the zone regularly to fulfill his needs in energy.

Boriska's parents, nice, educated and hospitable people, are worried about their son's fascinating talents. They do not know how others will perceive Boriska when he grows up. The say that they would be happy to consult an expert to know how to raise their wunderkind.

Being a doctor, his mother could not help but notice that the baby boy could hold his head already in 15 days after his birth. He uttered the first word 'baba' when he was four months old and started to pronounce simple words soon afterwards. At one year and a half he had no difficulties in reading newspaper headlines. At age of two years, he started drawing and leaned how to paint six months later. When he turned two, he started going to a local kindergarten. Tutors immediately noticed the unusual boy, his uncommon quickwittedness, language skills and unique memory.

However, his parents witnessed that Boriska acquired knowledge not only from the outer world, but through mysterious channels as well. They saw him reading unknown information from somewhere.

"No one has ever taught him," Boriska's mother said. "Sometimes he would sit in a lotus position and start telling us detailed facts about Mars, planetary systems and other civilizations, which really puzzled us," the woman said.

How may a little boy know such things? Space became the permanent theme of his stories when the boy turned two years old. Once he said that he used to live on Mars himself. He says that the planet is inhabited now too, although it lost its atmosphere after a mammoth catastrophe. The Martians live in underground cities, Boriska says.

The boy also says that, he used to fly to Earth for research purposes when he was a Martian. Moreover, he piloted a spaceship himself. It took place in the time of the Lemurian civilization. He speaks about the fall of Lemuria as if it occurred yesterday. He says that Lemurians died because they ceased to develop themselves spiritually and broke the unity of their planet.

When his mother brought him a book entitled "*Whom We Are Originated From"* by Ernest Muldashev, he got very excited about it. He spent a long time looking through the sketches of Lemurians, pictures of Tibetan pagodas, and then he told his parents of Lemurians and their culture for several hours non-stop. As he was talking, his mother noticed that Lemurians lived 70,000 years ago and they were nine meters tall… "How can you remember all this?" the woman asked her son. "Yes, I remember and nobody has told me that, I saw it," Boriska replied.

In Muldashev's second book "*In Search of the City of Gods"* he looked through pictures for a long time and recollected a lot about pyramids and shrines. Then he claimed that people would not find ancient knowledge under the Great Pyramid of Cheops. The knowledge will be found under another pyramid, which has not been discovered yet. "The human life will change when the Sphinx is opened, it has an opening mechanism somewhere behind the ear, I do not remember exactly," he said.

Boriska is one of so-called indigo children. They start to appear on Earth as a token of the forthcoming grand transformation of the planet.

"No, I have no fear of death, for we live eternally. There was a catastrophe on Mars where I lived. People like us still live there. There was a nuclear war between them. Everything burnt down. Only some of them survived.

They built shelters and created new weapons. All materials changed. Martians mostly breathe carbon dioxide. If they flew to our planet now, they would have to spend all the time standing next to pipes and breathing in fumes," Boriska said.

"If you are from Mars, do you need carbon dioxide?"

"If I am in this body, I breathe oxygen. But you know, it causes aging." Specialists asked the boy why man-made spacecraft often crash as they approach Mars. "Martians transmit special signals to destroy stations containing harmful radiation," Boriska replied.

The boy has deep knowledge of space and its dimensions. He is also aware of the structure of interplanetary UFOs. He talks about that like an expert, draws UFOs on slates and explains the way they work. Here is one of his stories: "It has six layers. The upper layer of solid metal accounts for 25 percent, the second layer of rubber – 30 percent, the third layer of metal – 30 percent, and the last layer with magnetic properties – 4 percent. If we give energy to the magnetic layer, spaceships will be able to fly across the Universe."

Boriska has a lot of difficulties with school. After an interview he was taken to the second grade, but soon they tried to get rid of him. He constantly interrupts teachers and says that they are wrong… now the boy has classes with a private tutor.

Translated by Julia Bulygina Pravda.ru

B. 12-Year-Old Genius Jacob Barnett Sets Out to Disprove Einstein

By Zachary Roth Yahoo News

In some ways, Jacob Barnett is just like any other 12-year-old kid. He plays Guitar Hero, shoots hoops with his friends, and has a platonic girlfriend.

But in other ways, he's a little different. Jake, who has an IQ of 170, began solving 5,000-piece jigsaw puzzles at the age of 3, not long after he'd been diagnosed with Asperger's syndrome, a mild form of autism. A few years later, he taught himself calculus, algebra, and geometry in two weeks. By 8, he had left high school, and is currently taking college-level advanced astrophysics classes—while tutoring his older classmates. The 12-year-old taught himself calculus, algebra and geometry in two weeks, and can solve up to 200 numbers of Pi. And he's being recruited for a paid researcher job by Indiana University.

Now, he's at work on a theory that challenges the Big Bang—the prevailing explanation among scientists for how the universe came about. It's not clear how developed it is, but experts say he's asking the right questions.

"The theory that he's working on involves several of the toughest problems in astrophysics and theoretical physics," Scott Tremaine of Princeton University's Institute for Advanced Studies—where Einstein (pictured) himself worked—wrote in an email to Jake's family. "Anyone who solves these will be in line for a Nobel Prize."

It's not clear where Jake got his gifts from. "Whenever I try talking about math with anyone in my family," he told the Indianapolis Star, "they just stare blankly."

But his parents encouraged his interests from the start. Once, they took him to the planetarium at Butler University. "We were in the crowd, just sitting, listening to this guy ask the crowd if anyone knew why the moons going around Mars were potato-shaped and not round," Jake's mother, Kristine Barnett, told the Star. "Jacob raised his hand and said,

'Excuse me, but what are the sizes of the moons around Mars?' "

After the lecturer answered, said Kristine, "Jacob looked at him and said the gravity of the planet ... is so large that (the moon's) gravity would not be able to pull it into a round shape."

"That entire building ... everyone was just looking at him, like, 'Who is this 3-year-old?'"

Chapter Six
Irrefutability of DNA Testing as 99.99% Accurate Shattered by Existences of Natural Human Chimeras

These cases of natural human chimeras challenge the blind faith which the scientific community places on the irrefutability of DNA testing. Forensic science cannot rely on DNA testing as the sole source of evidence, as it has done previously, as the criminal or victim may be a chimera. It follows that current maternity and paternity testing methods will have to be re-evaluated.

Even though there are two or more different sets of DNA in human chimeras, it may or may not be manifested as physical abnormalities. It may appear as phenotypic differences in eye colors, differential hair growth and coloring, "checkerboard" skin patterns, or missing or extraneous sexual organs. In 1998, at the University of Edinburgh, doctors examined a man who had complaints about an undescended left testicle. When they examined him, they were shocked to find an ovary and a fallopian tube in the male patient!

Most human chimeras, however, are not even aware of their conditions, because many of them appear completely normal. The most famous cases of chimerism to date are the linked cases of Lydia Fairchild (Year 2002) and Karen Keegan (Year 2000). Fairchild was pregnant with her third child, when she was separated from her partner, James Townsend. In order to obtain state welfare, she had to prove that she was the biological mother of her two born children. It was discovered, through DNA testing, that it was impossible that she was the biological mother of her two children because she bore no genetic similarity to them whatsoever. A case of welfare fraud ensued because the prosecutors believed the DNA results to be irrefutable. Even the testimony of Dr. Leonard Dreisbach, the obstetrician who had helped Fairchild give birth, did little to persuade the court in Fairchild's favor. The judge, perplexed by seemingly conflicting evidence, ordered that the third child, when born, to be tested as well. Surprisingly, the third child also showed no genetic similarities as well.

Fortunately for Fairchild, Karen Keegan also had similar experiences. Keegan needed a kidney transplant, and DNA testing for a compatible match with her two eldest sons showed that she had no genetic similarities to them at all. However, the doctors who worked with Keegan were familiar with the concept of chimerism and suggested that Keegan undergo further testing. Testing of her brothers and husband proved that her sons were related to them. Subsequent sampling of her skin and hair proved to be futile, but eventually matching DNA was found in her thyroid gland. It was the publication of this case, in the *New England Journal of Medicine*, which offered new insight on the case of Lydia Fairchild. Fairchild was found later on to be a chimera, with the second set of DNA found from her cervical smear. It was concluded that both Keegan and Fairchild were tetra gametic.

Chapter Seven
2000-Year-Old Analog Computer Found

The Antikythera mechanism is an ancient mechanical computer designed to calculate astronomical positions. It was recovered in 1900–1901 from the Antikythera wreck. Its significance and complexity were not understood until decades later. Its time of construction is now estimated between 150 and 100 BC. The degree of mechanical sophistication is comparable to a 19th century Swiss clock. Technological artifacts of similar complexity and workmanship did not reappear until the 14th century, when mechanical astronomical clocks were built in Europe. Jacques-Yves Cousteau visited the wreck for the last time in

1978, but found no additional remains of the Antikythera mechanism. Professor Michael Edmunds of Cardiff University who led the most recent study of the mechanism said: "This device is just extraordinary, the only thing of its kind. The design is beautiful, the astronomy is exactly right. The way the mechanics are designed just makes your jaw drop. Whoever has done this has done it extremely carefully ... in terms of historic and scarcity value, I have to regard this mechanism as being more valuable than the Mona Lisa."

The device is displayed at the National Archaeological Museum of Athens, accompanied by a reconstruction made and donated to the museum by Derek de Solla Price. Other reconstructions are on display at the American Computer Museum in Bozeman, Montana, the Computer History Museum in Mountain View, California, the Children's Museum of Manhattan in New York, and in Kassel, Germany.

Function

The device is remarkable for the level of miniaturization and for the complexity of its parts, which is comparable to that of 19th-century clocks. It has more than 30 gears, although Michael Wright has suggested as many as 72 gears, with teeth formed through equilateral triangles. When a date was entered via a crank (now lost), the mechanism calculated the position of the Sun and Moon or other astronomical information such as the locations of planets. Since the purpose was to position astronomical bodies with respect to the celestial sphere, with reference to the observer's position on the surface of the Earth, the device was based on the geocentric model.

The mechanism has three main dials, one on the front, and two on the back. The front dial has two concentric scales. The outer ring is marked off with the days of the 365-day Egyptian calendar, or the Sothic year, based on the Sothic cycle. Inside this, there is a second dial marked with the Greek signs of the Zodiac and divided into degrees. The calendar dial can be moved to compensate for the effect of the extra quarter day in the solar year (there are 365.2422 days per year) by turning the scale backwards one day every four years. Note that the Julian calendar, the first calendar of the region to contain leap years, was not introduced until about 46 BC, up to a century after the device was said to have been built.

The front dial probably carried at least three hands, one showing the date, and two others showing the positions of the Sun and the Moon. The Moon indicator is adjusted to show the first anomaly of the Moon's orbit. It is reasonable to suppose the Sun indicator had a similar adjustment, but any gearing for this mechanism (if it existed) has been lost. The front dial also includes a second mechanism with a spherical model of the Moon that displays the lunar phase.

There is reference in the inscriptions for the planets Mars and Venus, and it would have certainly been within the capabilities of the maker of this mechanism to include gearing to show their positions. There is some speculation that the mechanism may have had indicators for all the five planets known to the Greeks.

Source: http://en.wikipedia.org/wiki/Antikythera_mechanism

Chapter Eight
How did the Moon Form? Who Built the Moon?

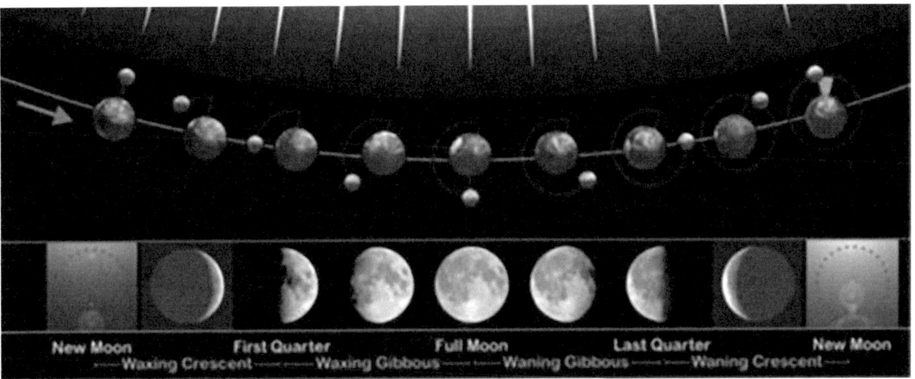

Not only is the Moon an apparently impossible object, we could not come to any other conclusion than the Moon is artificial. Is the Moon the creation of Intelligence?

The Moon is not only extremely odd in its construction, it also behaves in a way that is nothing less than miraculous. There are more than enough anomalies about the Moon to suggest it is not a naturally occurring body and was quite possibly engineered to sustain life on Earth.

The question of why the Moon had to be built is easy to answer. To produce life especially humans.

The Moon has some unique benefits for us humans. It has been nothing less than the incubator for life. If the Moon was not the exact mass, size and distance, that it has been at each stage of the Earth's evolution, there would be no intelligent life here. Scientists are agreed that we owe everything to the Moon. Without our Moon, the Earth would be as dead and solid as Venus, In the book 'Who built the Moon" that brought to light some extraordinary facts about the Moon, the authors Christopher Knight and Alan Butler claimed that:

- It is exactly 400 times smaller than the Sun, but 400 times closer to the Earth, so that both the Moon and the Sun appear to be precisely the same size in the sky, which gives us the phenomenon we call a total eclipse. Whilst we take this for granted, it has been called the biggest coincidence in the universe.
- Something has to maintain the Moon with its precise altitude, course and speed for it to maintain orbit.
- The Moon sits very close to the Earth, yet it is widely regarded as the strangest object in the known universe.
- The Moon does not have a solid core like every other planetary object. It is either hollow or has a very low intensity interior. Bizarrely, its concentration of mass are located at a series of points just under its surface which caused havoc with early lunar spacecraft.
- It acts as a stabilizer that holds our planet at just the right angle to produce the seasons and keep water liquid across most of the planet.
- The moon is unlike any satellite in our entire universe.

Because it is certain that it is 4.6 billion years old that raises some interesting points. The moon is older than the Earth by nearly 800,000 years according to scientific dating.

- Our Moon is the only moon in the solar system that has a stationary, near perfect circular orbit. It doesn't spin like a natural celestial body. Our Moon shares no characteristics of any moon within our Solar System.
- From any point on the surface of Earth, only one side of the Moon is visible.

Dr. Sean C. Solomon of the Massachussetts Institute of Technology said the lunar orbiter experiments had vastly improved knowledge of the Moon's gravitational field, and indicated the "frightening possibility that the moon might be hollow".

Isaac Asimov, a Russian professor of biochemistry, stated "We cannot help but come to the conclusion that the moon by rights ought not be there. The fact that it is, is one of those strokes of luck almost too good to accept. Small planets such as Earth, with weak gravitational fields, might well lack satellites…In general, then, when a planet does have satellites, those satellites are much smaller than the planet itself. Therefore, even if the Earth has a satellite, there would be every reason to suspect, that at best, it would be a tiny world, perhaps, 30 miles in diameter, but that is not so. Earth not only has a satellite, but it is a giant satellite, 2160 miles in diameter."

Some lunar rocks have been found to contain 10 times more titanium than titanium rich rocks on Earth. Titanium is used in supersonic jets, deep diving submarines and spacecraft. Dr Harold Urey, a winner of the Nobel Prize for Chemistry, said he was "terribly puzzled by the rocks from the Moon, and in particular their Titanium content'. He said the samples were "mind-blowers", and that he could not account for the Titanium.

Irwin Shapiro from the Harvard-Smithsonian Center for astrophysics said, "The best explanation for the Moon is observational error---the Moon does not exist".

Robin Brett, a NASA scientist stated, "It seems easier to explain the non-existence of the Moon than its existence."

In November 1969, NASA intentionally crashed a lunar module causing an impact equivalent to one ton of TNT. The shockwaves built up and NASA scientists said the Moon "rang like a bell". The reverberation from it continued for 30 minutes.

Ken Johnson, a supervisor of the data and photo control department during the Apollo Missions said that the Moon not only rang like a bell, but the whole Moon "wobbled" in such a precise way, that it was "almost as though it had a gigantic hydraulic damper strut inside it.

In 1970, Mikhail Vasin and Alexander Shcerbakov from Soviet Academy of Sciences, produced an article for Sputnik magazine called "Is the Moon the creation of Alien Intelligence?" The article contained the following: The outer surface of the Moon is extremely hard and contains minerals like Titanium. Moon rocks have been found to contain processed metals, including Brass and Mica, and the elements Uranium 236 and Neptunium 237 that have never been found to occur naturally. Uranium 236 is a long lived radioactive nuclear waste and is found in spent nuclear and reprocessed Uranium. Neptunium 237 is a radioactive metallic element and a by product of nuclear reactors and the production of Plutonium. If a material had to be devised to protect a giant artificial satellite from the unfavorable effects of temperature, from cosmic radiation and meteorite bombardment, the experts would probably have hit upon these refractory metals. In that case, it is not clear why lunar rock is such an extraordinary poor heat conductor—a factor which so amazed astronauts? Wasn't that what the designers of this super Sputnik of the Earth were after? **From the engineer's point of view, this spaceship of ages long past which we call the Moon is superbly constructed".**

Part 5
New Mentality Brings About New Discoveries

In Man's conquest of progress, his own prejudice and ignorance must be the first to surrender. From there, he will step his way across the heavens to the edge of infinity. No doubt he will stumble and fumble. Each step will be as uncertain as the last, yet each will bring him closer to ultimate truth. Throughout history, a handful of brave scientists and technicians pave the way to the future. Their mission: to collect information that will eventually enable man to step into the better world of tomorrow, to use their present knowhow as a springboard to the brave new world that they dare not imagine. They make their slow, uncharted way across the waters of mysticism, pseudoscience, and the paranormal of their times. The water's surface has depths and dangers that are, as yet, unprobed . . .

It is with these people that our new world has come to being, mapped, chartered, and indexed . . .

If we accept the tenuous fragility of modern progress through trial and error, misfortunes and sacrifice, perhaps we can begin the second dawning of humankind . . . today.

Chapter One
Monoatomic Gold as Superconductor

http://www.halexandria.org/dward479.htm

Dan Sewell Ward

A Monoatomic form of the element—in which each single atom is chemically inert and no longer possesses normal metallic characteristics; and in fact, may exhibit extraordinary properties. The atom's intrinsic temperature is now about one oK or close enough to absolute zero that superconductivity is a virtually automatic condition.

A case in point is gold. Normally a yellow metal with a precise electrical conductivity and other metallic characteristics, the metallic nature of gold begins to change as the individual gold atoms form chemical combinations of increasingly small numbers. At a micro cluster stage, there might be thirteen atoms of gold in a single combination. Then, dramatically, at the monoatomic state, gold becomes a forest green color, with a distinctly different chemistry. Its electrical conductivity goes to zero even as its potential for superconductivity becomes maximized. Monoatomic gold can exhibit substantial variations in weight, as if it were no longer fully extant in space-time.

Other elements which have many of these same properties are the precious metals, which include ruthenium, rhodium, palladium, silver, osmium, iridium, platinum, and gold. All of these elements have, to a greater or lesser degree, the same progression as gold does in continuously reducing the number of atoms chemically connected. Many of these precious elements are found in the same ore deposits and in their monoatomic form are often referred to as the white powder of gold. Monoatomic elements apparently exist in nature in abundance. Precious metal ores are, however, not always assayed so as to identify them as such. Gold miners, for example, have found what they termed "ghost gold"—"stuff" that has the same chemistries as gold, but which were not yellow, did not exhibit normal electrical conductivity, and were not identifiable with ordinary emission spectroscopy. Thus, they were more trouble than they were worth, and generally discounted.

The mining activity of what is considered the best deposit in the world for six of these elements (Pd, Pt, Os, Ru, Ir, and Rh) yields one third of one ounce of all these precious metals per ton of ore. But this is based on the standard spectroscopic analysis. The distinguishing characteristic between the first and second readings of the emission spectroscopy for the precious metals is that all of them come in two basic forms. The first is the traditional form of metals, yellow gold, for example. The second is the very nontraditional form of the metal, the monoatomic state. The chemistries and physics of these two different states of these metals are radically different. More importantly, when the atoms are in the monoatomic state, things really begin to get interesting!

A key to understanding monoatomic elements is to recognize that the monoatomic state results in a rearrangement of the electronic and nuclear orbits within the atom itself. This is the derivation of the term: Orbitally-Rearranged Monoatomic Element (ORME).

A monoatomic state implies a situation where an atom is "free from the influence of other atoms." Is this, perhaps, a violation of some very basic, absolutely fundamental law of the universe—which says that nothing is separate? If such a law constituted reality, then a necessary condition for monoatomic elements to even exist would require them to be superconductive, just in order to link them through all distance and time to other superconducting monoatomic elements. This would be necessary in order to prevent separation.

Chapter Two
Anticipation of Discovery in Astronomy Always Leads to New Discoveries

It was perturbations in Uranus's motion that led to the discovery of Neptune.

Astronomers had noticed increasing errors in their models for the orbit of Uranus and eventually ruled out many explanations besides an outer planet that could be perturbing its orbit.

In 1846, the planet Neptune was discovered after its existence was predicted.

This precise prediction of the new planet and its location was striking confirmation of the power of Newton's theory of gravitation.

Later, similar calculations on supposed perturbations of the orbits of Uranus and Neptune suggested the presence of yet another planet beyond the orbit of Neptune. Eventually, in 1930, a new planet, Pluto, was discovered.

We have even more amazing discoveries recently:

A trans-Neptunian object is any object in the solar system that orbits the sun at a greater average distance than Neptune.

Sedna

90377 Sedna is a large trans-Neptunian object, which, as of 2012, was about three times as far from the sun as Neptune. For most of its orbit, it is even farther from the sun than at present, with its aphelion estimated at 937 astronomical units, making it one of the most distant known objects in the solar system other than long period comets. Sedna's exceptionally long and elongated orbit, taking approximately 11,400 years to complete, and distant point of closest approach to the sun, at 76 AU, have led to much speculation as to its origin.

Eris, Pluto, Makemake, and Haumea

As of November 2009, two hundred trans-Neptunian objects have their orbits well enough determined that they have been given a permanent minor planet designation.

The largest known trans-Neptunian objects are Eris and Pluto, followed by Makemake and Haumea. The Kuiper belt, scattered disk, and Oort cloud are three conventional divisions of this volume of space.

Comet Lovejoy

Comet Lovejoy is a long period comet and Kreutz Sungrazer. It was discovered in November 2011 by Australian amateur astronomer Terry Lovejoy. The comet's perihelion took it through the sun's corona on 16 December 2011, after which it emerged intact and continued on its orbit to the outer solar system. In the process, it astounded all scientists.

Comet 96P/Machholz 1

Comet 96P/Machholz last came to perihelion on July 14, 2012 and will next come to perihelion on October 27, 2017. 96P/Machholz has an estimated radius of around 3.2 km.

Machholz 1 is unusual among comets in several respects. Its highly eccentric 5.2-year orbit has the smallest perihelion distance known among numbered/ regular short period comets, bringing it considerably closer to the sun than the orbit of Mercury.

Chapter Three
God Particle Having Been in Existence for Millenia, Found in July 2012

http://www.technewsworld.com/story/77540.html

By Richard Adhikari TechNewsWorld 03/15/13

"There is no indication that this is not a Standard Model Higgs at present," Dan Green, a senior scientist emeritus at Fermilab's CMS Center, told TechNewsWorld. "More data has shrunk the possibilities of a non-Standard Model Higgs."

A large decay rate into two photons was seen earlier, but that had large statistical errors, and the data "has now settled to a value closer to the Standard Model," he said.

Next Steps

The issue scientists are facing now is finding out what protects the Higgs boson's low mass from radiative corrections. "Something limits the mass to about 125 GeV and we don't know what that is—yet," Green said.

There's a level of certainty that the particle discovered is indeed the Higgs boson, but "in science, you cannot say with certainty what you have seen," he said. The scientists want what they refer to as "5-sigma" confidence "for all the possible decay modes of the Higgs, and that will take several years of the LHC running at increased luminosity and higher energy."

The scientists are "being super-cautious," William Newman, professor of Earth and space sciences at the University of California at Los Angeles, told TechNewsWorld. "The original discovery was a 5-sigma discovery and that's a super-gold standard; it's almost the holy grail."

Validation of the standard model at this level "excludes a large variety of competing models that people have been putting forth, and that dramatically constrains how the universe can behave," he said.

Why Search for Decay?

The Higgs boson has no spin, no electric charge and no color charge, and is its own antiparticle. It doesn't exist except as a theoretical construct, and here's why: It accompanies the Higgs Field, a possible invisible field of energy that exists throughout the universe.

In the standard model, the Higgs Field consists of four components: two charged component fields and two neutral ones. Both the charged components and one of the neutral fields are Goldstone bosons. The quantum of the last, neutral component is theoretically realized as the Higgs boson.

Quantum mechanics theory holds that particles will decay into a set of lighter particles if it's possible for them to do so. The likelihood of this occurring depends on various factors, including the difference in mass between the original particle, and the ones it decays into, and the strength of the interactions. The standard model fixes most of these factors except for the mass of the Higgs boson itself.

Chapter Four
Galileo, Your Best Bet Is to Be Born after Newton

Sir Isaac Newton, born on December 25, 1642, was an English physicist, mathematician, astronomer, natural philosopher, alchemist, and theologian who has been considered by many to be the greatest and most influential scientist who ever lived.

Galileo Galilei, born on February 15, 1564, was an Italian scientist who supported Copernicanism, the idea that Earth orbits the sun. Galileo defended his views in *Dialogue Concerning the Two Chief World Systems*. For doing so, he was tried by the Roman Inquisition, was found "suspect of heresy," and spent the rest of his life under house arrest. His findings changed our world view for all time. Galileo died a natural death in the year 1642. He was seventy-seven years old and was serving his life imprisonment sentence in Arcetri when he died.

In fact, Newton did not discover gravity. It was actually Galileo. Galileo dropped different weight objects from the Tower of Pisa and noted they all fell at the same rate, and he did many other experiments along these lines.

In fact, Galileo died a year before Newton was born.

What if Galileo Galilei were born after Sir Isaac Newton?

Galileo Galilei would have had an easier life if he were born after Sir Isaac Newton. With the discovery of gravitational pull, it could be demonstrated easily that the Earth rotated around the Sun. And he wouldn't have spent the rest of his life under house arrest.

Sometimes, some truths or discoveries come too early, especially for people who are indifferent, arrogant, and ignorant. Politics, greed, and an irrational resistance to change have prompted people to denounce the latest discovery or technical innovation as rubbish, often as a knee-jerk reaction. Egos, vested interests, politics, moral and religious objections, and evasion of responsibility are all underlying factors to the fact that many beneficial discoveries have been swept under the carpet.

Many, many lives that could be saved with the latest discoveries and inventions have been lost with the latest findings suppressed and the inventor dying unheralded. Inventors often need a great deal of tenacity, as well as vision, to overcome the skepticism and ridicule that come as an avalanche. Sometimes discoverers and inventors are persecuted. A breakthrough would have far-reaching consequences for the entire religious sector based on Galileo Galilei's work—which, in itself, may be the problem. Until a very long time after his death, his books were banned and his writings and findings were considered deviant, false and of the devil.

Now who is to say what discoveries should come earlier than others?

Conclusion

Sitchin may be a hoax in some people's eyes. We might not value what he has to offer us now. In time, when other scientific discoveries are made, they may exonerate Sitchin. To Sitchin's critics, who would like to see his books banned and his writings and findings considered deviant and false, they really need to appreciate how Sitchin's findings have changed our world view for all time.

Chapter Five
The Greeks Knew the Earth Was Round and Measured Its Circumference Before Copernicus and Galileo

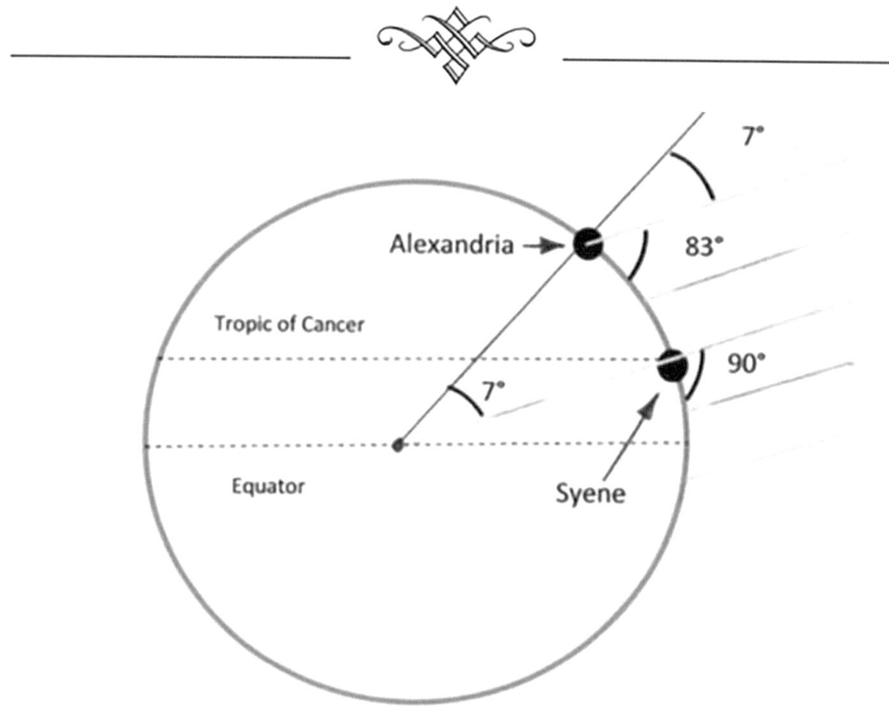

The ancient Greeks knew a thing or two. Nowadays we say Copernicus discovered the Earth was round, for example, but the Greeks had that figured out almost two thousand years earlier!

How Did They Know the Earth Was Round?

The common Greek knew the Earth was round back then because Greeks were sailors. As a ship would go away from a port they noticed that the sails of the ship would disappear last. Thus, they knew that the Earth has a curvature and was not flat!

Now remember that all the stars seem to move in a big circle over our heads because the Earth itself is spinning. The North Star is the center of the circle, and stays put.

However, Aristotle noticed that when you are down south in Egypt, the North Star is close to the horizon. When you are up north in Greece again, it's high in the sky.

How is that possible, since the North Star is supposed to stay put? It's possible, Aristotle concluded, only if the Earth is round.

Think of it this way . . . When you are standing at the north pole, where is the North Star? Directly overhead. Now place yourself at the equator. Where is the North Star? Yep—it's down on the horizon, because you have traveled along a curve. Go even further south and you won't be able to see it at all; the Earth will hide it. Go further north and it seems to rise higher and higher.

What about Measuring the Earth's Circumference?

Contrary to popular belief, Christopher Columbus did not discover that the Earth is round. Eratosthenes (276–194 B.C.) made that discovery about seventeen hundred years before Columbus!

He calculated the circumference of the Earth without leaving Egypt.

Eratosthenes was a Greek mathematician, astronomer, and the father of geography with the invention of a system of latitude and longitude.

How did he do it?

1. Eratosthenes first measured the distance between Alexandria and Syene to be five hundred miles. This is based on the estimated average speed of a caravan of camels that traveled this distance. Camels traveled the distance many times to obtain an average estimate of the distance.

2. Eratosthenes observed that at noon on the summer solstice, the longest day of the year, no shadow was cast in a well in the city of Syene in Egypt.

3. At that time the sun was directly overhead. He then assumed that the sun was so far off that its rays hit the Earth in parallel.

4. He further assumed that the Earth was shaped like a ball from the stories of Greek sailors.
5. He surveyed the city of Syene to be on the same meridian as the city of Alexandria. because the sun set at the same time when it was directly between the two cities.
6. So, on the summer solstice, at noon, in Alexandria, Eratosthenes measured the angle of the sun's rays to be 7.2 degrees angle with a tall rod.
7. He also knew, from measurement, that in his hometown of Alexandria the angle of elevation of the sun was 1/50th of a circle (7°12') south of the zenith on the solstice noon.
8. Assuming that the Earth was spherical (360°), and that Alexandria was due north of Syene, he concluded that the meridian arc distance from Alexandria to Syene must therefore be 1/50 = 7°12'/360°, and was therefore 1/50 of the total circumference of the Earth.
9. Since he knew that five hundred miles was 1/50 of the circumference of the Earth, he needed to use fifty of these lengths to surround the Earth.
10. So he multiplied five hundred miles by fifty to get twenty-five thousand miles.
11. Then he added two hundred miles more to make up for what he thought were bad measurements.

So, he calculated the circumference of the Earth to be 25, 200 miles.

That measurement is close to the actual 24,901 miles. In those days they couldn't easily make very accurate measurements of the distance between two places, so this could cause a lot of error.

Chapter Six
"Fellow Roman Citizens . . . Rome Is Not the Only Mighty Empire in the World!"

"You called yourself the greatest empire in existence. Do you believe there exists an empire called the Chinese Empire, which is at least as large if not bigger than the Roman Empire?"

"You are most proud of your architectural achievements. Do you know there are pyramids that are very much harder to construct?"

"Other than the Romans and barbarians, do you know of the existence of the Eskimos, American Indians, Australian Aborigines . . .?"

"Do you believe your Roman scholars and authorities know most of everything? Do you have many financial crises? Social Inequalities?"

"Would you believe in some strange lands, strange people, strange ideas, and strange discoveries if your Roman scholars and authorities told you not to?"

"Do you know Korea? The country has been in existence for thousands of years!"

Well, the last question is too hard. Many Americans did not even know Korea existed until the American government decided to send troops to Korea in the Korean War, despite the fact that the country has been in existence for thousands of years!

With these facts in mind, how are we even sure on what we know, or for that matter, what we don't know?

Conclusion

Sitchin was trying hard to tell us something we don't know.

Chapter Seven
Conversation with the Wright Brothers in 1902, Just before Their Important Breakthrough

If we go back to 1902, with all the knowledge we have about flying and aerodynamics, will we find any person who can be convinced that flying is possible, let alone who can be convinced that man would be on the brink of flying through the air at the speed of sound? Can we even convince the Wright Brothers? Maybe not.

Wright Brothers: Well, Mr. Flying, if your motorized craft can go faster than the speed of sound, why isn't your flimsy craft torn to pieces by the wind? Besides, the human body cannot take strong acceleration, according to the laws of nature.

Mr. Flying: It is possible to pass the sound barrier by designing the wings and body to move the shock wave down the plane as you surpass the speed of sound.

Wright Brothers: How do you make wings with enough lifting power? How do you avoid the airplane spinning out of control by the wing warping mechanism?

Mr. Flying: Combinations of aluminum, magnesium, and small amounts of copper and manganese make a light but strong alloy called duralumin, which is suitable for airplane parts.

Wright Brothers: Really? Planes made of alloy? Well, if your plane can fly faster than sound, then why don't you just fly to the moon?

Mr. Flying: Trained astronauts in a space shuttle can do that, not an airplane.

Wright Brothers: I see. Fly to the moon in a space shuttle? Man will probably not fly in their lifetime. Not in an airplane or space shuttle. Is that a fantasy?

Mr. Flying: Perhaps, perhaps not.

Chapter Eight
Suppressed Discoveries Become Forbidden and Then Forgotten

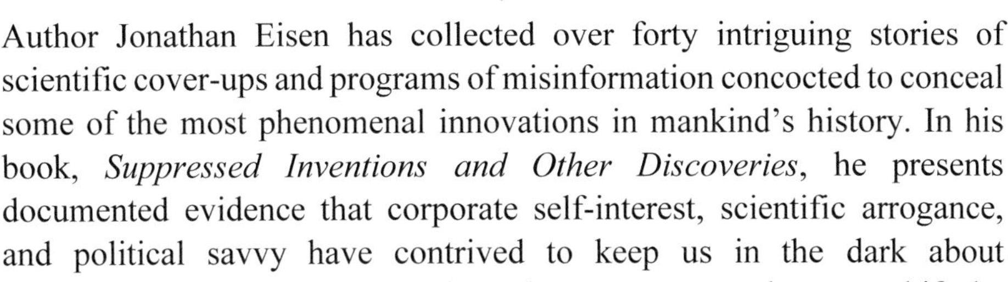

Author Jonathan Eisen has collected over forty intriguing stories of scientific cover-ups and programs of misinformation concocted to conceal some of the most phenomenal innovations in mankind's history. In his book, *Suppressed Inventions and Other Discoveries*, he presents documented evidence that corporate self-interest, scientific arrogance, and political savvy have contrived to keep us in the dark about technological breakthroughs or interplanetary contact that may shift the current balance of power.

Area I

- Alternative Medical Therapies
- The Great Fluoridation Hoax
- Deadly Mercury—How It Became Your Dentist's Darling
- The Alzheimer's Cover-Up
- Vaccinations—Adverse Reactions Cover-Up
- AIDS and Ebola—Where Did They Really Come From?
- Polio Vaccines and the Origin of AIDS
- Oxygen Therapies, the Virus Destroyers
- Oxygen Therapy
- The FDA
- Harry Hoxsey—The AMA's Successful Attempt to Suppress My Cure for Cancer
- Royal Raymond Rife and the Cancer Cure That Worked!
- The Persecution and Trial of Gaston Naessens

- Dr. Max Gerson's Nutritional Therapy for Cancer and Other Diseases

Area II

- The Suppression of Unorthodox Science
- Science as Credo
- Sigmund Freud and the Cover-Up of "The Aetiology of Hysteria"
- The Burial of Living Technology
- Egyptian History and Cosmic Catastrophe
- Archaeological Cover-Ups?
- Introduction to Bread From Stones
- Scientist With an Attitude—Wilhelm Reich
- The AMA's Charge on the Light Brigade
- The Neurophone

Area III

- The Suppression of UFO Technologies and Extraterrestrial Contact
- Breakthrough as Boffins Beat Gravity
- Antigravity on the Rocks—The T. T. Brown Story
- Did NASA Sabotage Its Own Space Capsule?
- Extraterrestrial Exposure Law Already Passed by Congress
- The Stonewalling of High Strangeness
- UFOs and the U.S. Air Force
- UFOs and the CIA—Anatomy of a Cover-up
- NASA
- UFO Phenomena and the Self-Censorship of Science
- Mars—The Telescopic Evidence

- Never a Straight Answer—A Book Review of NASA Mooned America

Area IV

- The Suppression of Fuel Savers and Alternate Energy Resources
- Nikola Tesla—A Brief Introduction
- Tesla's Controversial Life and Death
- Transmission of Electrical Energy without Wires
- From the Archives of Lester J. Hendershot
- Gunfire in the Laboratory—T. Henry Moray and the Free Energy Machine
- Sunbeams from Cucumbers
- Archie Blue
- The Story of Francisco Pacheco
- Amazing Locomotion and Energy Super Technology and Carburetors
- The Charles Pogue Story
- News Clips on Suppressed Fuel Savers

Chapter Nine
The Elephant of Reality: God, Evolution, and Intelligent Design

Chronology of Human Events and the Rise of Different Theories of Human Origins 4.5 Billion Years Ago

Our sun created the planet Tiamet, the proto-Earth. Tiamat orbited the Sun counterclockwise.

Two Hundred Million Years Ago

Panspermia started two hundred million years ago when Nibiru's moon and Nibiru hit Tiamat and left no crust at all in the Pacific Gap, only a gaping hole. Initial life-forms were transferred to New Earth.

Evolution slowly picked up pace as conditions changed for the better to favor the emergence of life-forms.

Four Hundred and Fifty Thousand Years Ago

Annunaki humanoids arrived on planet Earth, setting up Nibiru Colony

Four Hundred Thousand Years Ago

The Annunaki started **non-intelligent design.** They created many hybrid species and chimeras out of genetic errors. They also created many abhorrent types of experimental humanoids and animals.

Three Hundred Thousand Years Ago

The Annunaki started **intelligent design.** They created the perfected specimens of Adam and Eve as human progenitors to be used as mining workers.

Two Hundred and Fifty Thousand Years Ago

Humans multiplied in great numbers. For effective rule of the human population, the Annunaki humanoids taught the humans to revere them as **gods.**

12,000 B.C.

The Annunaki strengthened their lordship over the humans. Various structures for worshipping them as **gods** were erected. Treating humans as domesticated animals, they brought new species of plants and animals from their home planet Nibiru as rewards for humans who labored in the mines. They also domesticated wild plants and animals native to Earth for humans as food and other consumption.

9,000 B.C

Intermarrying earthlings, especially females, further hybridized the human genome. Men of renown known as heros or nephilims further sped up **human evolution.** Human beings now seemed to be perfectly designed with a purpose. This prompted posterity to term their birth through "**intelligent design.**"

6,000 B.C.

The Anunnaki created bloodlines to rule humanity on their behalf, and these are the families still in control of the world to this day. Kingship was granted to humanity by the Anunnaki, and it was originally known as Anuship after An or Anu, the ruler of the 'gods.' Eventually, kingships became prevalent all over the globe. The royal families and aristocracy of Europe, Asia, and the Middle East are obvious examples of this. Most of the world's organized religions were also established at around this time.

5,000 B.C.

The constant power struggle and the quest for domination by the Annunaki led a minority of humans to escape the control of the Annunaki by migrating to other areas. Some of these became nonbelievers of the Annunaki. Their teachings were passed down to posterity. Some of their descendants became **agnostics and atheists.**

1,831 A.D.

"The Voyage of the Beagle" refers to the second survey expedition of the ship HMS Beagle, which set sail from Plymouth Sound on 27 December 1831 under the command of Captain Robert FitzRoy, R.N. Darwin's notes made during the voyage included comments illustrating his changing views at a time when he was developing his theory of Evolution by Natural Selection and included some suggestions of his ideas, particularly in the second edition of 1845.

1855 A.D.

In February 1855, while working in Sarawak on the island of Borneo, Alfred Russel Wallace wrote "On the Law which has Regulated the Introduction of New Species", a paper which was published in the Annals and Magazine of Natural History in September 1855. By February 1858, Wallace had been convinced by his biogeographical research in the Malay Archipelago of the reality of evolution.

1859 A.D.

On November 1859, Charles Darwin published "On the Origin of Species", a work of scientific literature by which is considered to be the foundation of evolutionary biology.

2013 A.D.

History replays itself over and over again in man's history of conflicts, conquests, and conjectures. Man has been warring one another as individuals and as nations. All theories of human origins seek to dominate to the exclusion of other theories.

Part 6

Pursue Truth Aggressively—Essentially and Especially your own Truth!

We have begun the space age. We are all trying to maneuver through a brand-new, unfamiliar landscape using an old map drawn up by religion and Darwin. We need a new guiding principle.

Unaware of the new realities, we are like the last remaining jungle fighters of Japan still hiding in the swamps blindly unaware that the war was over decades ago.

"The past history of humanity shows us that each stage of its development necessitates an uprooting and renewing of fundamental beliefs in our scientific, social, philosophic, and religious conceptions."—M. Plank. Our history is also full of silly mistakes that must make us smile today. Just like new things that happen everyday, another new theory will soon appear, and the new theory that just appeared will be replaced by a newer one, which itself will be replaced by a better one, and so on ad infinitum . . . Our history is also full of icons who also made silly mistakes.

Ernest Rutherford implied that energy produced by an atom was insignificant and anyone who expected to find a new source of energy in that transformation was dreaming. Even Albert Einstein indicated that there was no indication that nuclear energy would be accessible one day, and yet Hiroshima happened twelve years later.

Author Arthur C. Clarke believed that when even eminent and distinguished scientists say something is possible, he could be right, but if he says it's impossible, then there's a good chance he's wrong.

Sitchin's work will make us dream impossible dreams that we have not dreamed before. With that, he will bring us to new progress that we cannot even dream of, if we understand our past and then foresee our future. It is a new reality too incredible to accept, too awesome to face. The establishment of a new paradigm need not cause the fall of the old paradigm. The time has come to shift into a higher gear and reveal what the future holds for us by examining our very own past.

God particles are around for billions of years but we only know they exist in 2012!

Over the years, startling evidence has been uncovered, challenging established notions of the origins of life on Earth—evidence that suggests the existence of an advanced group of extraterrestrials who once inhabited our world.

Many researchers have since uncovered incredible findings having depth, complexity, and far-reaching effects in support of Sitchin. This book is an attempt to bring about the general awareness in the public of a topic so important and so controversial. The reader is encouraged to dig deeper into all the material, the artifacts, the "what if," and reach his or her own conclusions.

We need to be aware of our own tunnel vision imposed by the current scientific and religious orthodoxy. "Your time is limited, so don't waste it living someone else's life. Don't be trapped by dogma—which is living with the results of other people's thinking. Don't let the noise of others' opinions drown out your own inner voice. And most important, have the courage to follow your heart and intuition. They somehow already know what you truly want to become. Everything else is secondary" (Stanford commencement speech, June 2005, Steve Jobs).

Wake up! This is not science fiction. Truths that were yet unknown to this day have been unveiled.

The highest form of ignorance is when you reject something you don't know anything about.

Wayne Dyer

"What we call imagination is actually the universal library of what's real. You couldn't imagine it if it weren't real somewhere, sometime."

- Terence Mckenna -

Part 7
What we used to believe

Chapter One
Inertia, Indifference and Ignorance

It is evident in our history that scientific and technological progress has been retarded by authoritative persons who have declared innovations and inventions be either wrong or of no practical use. Many pioneers for discovery had spent their lifetimes trying to convince others on what they knew or found only to fall upon death ears most of the times. Even after the preponderance of evidences were raised, people could still be in denial.

Most forefront thinkers or scientist were first ridiculed, and then vehemently opposed with strong support from the masses or ignorant populace using (insufficient) logical reasoning, (incomplete) careful analylsis and (inexhaustive) evidence. After costly mistakes were made, progress was impeded, situation worsened, then people turned back to those supposedly outrageous claims and concluded that the evidence was too much to not acknowledge. From then on, the idea or discovery was thought as self-evident truth.

Let us not forget the following facts:

For centuries most societies believe the earth is flat. All the authorities and leaders in various disciplines from most cultures were the guardians of this faith. China held this believe as late as the 17 centuries.

We are the same people who used to believe that Monster lived at the edge of the earth.

We are the same people who used to believe spontaneous generation of life was possible --until until Luis Pasteur's experiment in the 19 century.

We are the same people who used to believe the earth was immersed in luminous ether, and then it was discovered that light waves propagate through space.

We are the same people who used to believe flying was not possible until December 17, 1903 When the Wright Brothers did it against the numerous critics.

Neutrons are around for billions of years but we only know they in 1932!!!

In 1970 Plate Techtonics replaced our old belief that earth was expanding and causing the fault lines, First Asteroid Companion of Earth 2010 TK7 was only discovered in 2010.

I think its evident that times are changing and while some things were dead on, some were just wanting to be blown open with new discoveries.

The discoveries made with Sumerian Clay Tablets is well suited to anyone, professionals and amateurs, science buffs and historians, pioneers and followers. Depending on the extent of your engagement, it can be overwhelming or it can be entertaining. It can lead to the Grand Truth, or it can lead to elements of Fringe with new discoveries made directly, accidentally or even somewhere else not at all related.

One the one hand we have the Impossibilist Camp who decry foul at every opportunity. On the other hand we have authoritative persons who throughout history have labeled all innovations and inventions as Fringe Sciences unworthy of serious investigation by main-stream science.

The irrational resistance to change and the desperate attempt to cling to our safety shell, lies in our inherent arrogance and ignorance, sense and sensibility. It is said that our civilization has make progress in leaps and bounds. It can also be said that misjudgment and ignorance have impeded and prevented progress be made in leaps and bounds.

Who should be made into the laughing stock ? The Clown ? The Comedian ? The Scholar ? The Scientist ? Ourselves ? Many inventors died unheralded. Many succumbed to the scepticism and ridicule that were piled upon them from all kinds of directions.

Many Armchair Critics are marked by cynicism and the manipulation of data to fit their personal beliefs. Scientific Theories that were once right and mighty had been proven wrong later. This is the case even with the best and most thorough scientific scrutiny and investigation. There is no knowing when a better Scientific Theory will come out to replace the old.

It is clear that the Sumerian Clay Tablets Study will continue to interest areas of science, history, archaeology, cosmology etc. etc. due to its continued "effect" of encompassing studies from unrelated fields.

What might be the beauty of Sitchin's Theory is that it gives people, scientists and scholars the confidence to create new theories and discover new findings even though the final and conclusive proof may still be unknown or unaccepted by the populace or scientific mainstream.

Chapter Two
Conversation with the Wright Brothers in 1902

If we go back to 1902, with all the knowledge we have about flying and aerodynamics, will we find any person who can be convinced that flying is possible? Let alone convincing them that man would be on the brink of flying through the air at the speed of sound? Can we even convince the Wright Brothers? Maybe not.

Wright Brothers: Well Mr. Flying, if your motorized craft can go faster than the speed of sound, why isn't your flimsy craft torn to pieces by the wind? Besides, the human body cannot take strong acceleration, against law of nature.

MR. Flying: It is possible to pass the sound barrier by designing the wings and body to move the shock wave down the plane as you surpass the speed of sound.

Wright Brothers: How do you make wings with enough lifting power? How do you avoid the airplane to spin out of control by the wing warping mechanism?

MR. Flying: Combinations of aluminum, magnesium, and small amounts of copper and manganese make a light but strong alloy called duralumin which is suitable for airplane parts.

Wright Brothers: Really? Planes made of alloy? Well, if your plane can fly faster than sound then why don't you just fly to the moon?

MR. Flying : Trained Astronauts in a space shuttle can do that. Not an airplane.

Wright Brothers: I see. Fly to the moon in a space shuttle? Man will probably not fly in their lifetime. Not in an airplane or space shuttle. Is that a fantasy?

MR. Flying: Perhaps, perhaps not.

Chapter Three
Physical Center of the Universe and Biological Center of the Universe

Geocentricity is the belief and teaching that the earth is the center of the solar system and/or the universe.

Because we have huge egos, we believe everything revolves around us until 500 years ago.

Because human vanity knows no limits, we want to believe we are the epitome of developed species, the most advanced, intelligent, and civilized. That is, until The Human Genome Project shocked us with some real surprises.

The first surprise was the vast percentage—at least 97 percent—of the human DNA is inactive, thus termed "junk DNA."

The second surprise was that even though we seem to have the longest DNA molecule among all other species, we use the smallest fraction of it in relation to the other species. Both of these issues raise the question of efficiency of our DNA evolution.

We would expect that humans should end up having most genes, but strangely, this is not the case.

Species	Number of genes
rice	63,000,
grapes	30,000
water flea	31,000
sea urchin	23,300
humans	23,000
dogs	19,300
chicken	16,700
fruit fly	14,800

We would also expect that humans should end up having the biggest genome, but strangely, this is not the case either.

Species	Genome Size in Picograms of DNA
Paris japonica, native plant of Japan	152.23
Marbled lungfish	132.83
Human	3.0
Encephalitozoon intestinalis parasite	0.03

The time has again come for us to dispel our mystic belief that we are the ultimate prize of Creation, evolution, or intelligent design.

We can then fairly say, "Genome size does not correlate with evolutionary status, nor is the number of genes proportionate with genome size."

Then, what are the consequences of such a large genome, and does it really matters if one organism has more DNA than another?

The consequences operate at all levels, from the cell up to the whole organism, and beyond. In plants, research has demonstrated that those with large genomes are at greater risk of extinction, are less adapted to living in polluted soils, and are less able to tolerate extreme environmental conditions—all highly relevant in today's changing world.

Another example of the significance and importance of genome size in both animals and plants is the fact that the more DNA there is in a genome, the longer it takes for a cell to copy all its DNA and divide. The knock-on effect of this is that it can take longer for an organism with a larger genome to complete its life cycle than one with a small genome. It is no coincidence that living in deserts, many plants that must grow quickly after rains have small genomes, enabling them to grow rapidly. In contrast, species with large genomes grow much more slowly and are excluded from such habitats.

Genome size is also positively correlated with nuclear size (The more DNA you have, the more space you need for it.), and, in many cases, also with cell size, which can have knock-on consequences at the whole organism level.

http://www.sciencedaily.com/releases/2010/10/101007120641.htm

Prior to the current genomic era, it was suggested that the number of protein-coding genes that an organism made use of was a valid measure of its complexity. It is now clear, however, that major incongruities exist and that there is only a weak relationship between biological complexity and the number of protein coding genes. For example, using the protein-coding gene number as a basis for evaluating biological complexity would make urochordates and insects less complex than nematodes, and humans less complex than rice. Increasing biological complexity is positively correlated with the relative genome wide expansion of nonprotein-coding DNA sequences.

Ryan J. Taft and John S. Mattick

Rowe Program in Genetics, Department of Biological Chemistry, University of California, Davis, School of Medicine; ARC Special Research Centre for Functional and Applied Genomics, Institute for Molecular Bioscience, University of Queensland.

Chapter Four
History Is Wrong

History is collective memory; yet, even our own memory errs at times, and no real memory extends beyond three generations. There are written sources, but each one of those might easily prove a forgery. There are material remnants of archaeological nature, but they may be misdated and misinterpreted.

The extended history of man living in caves, inventing tools, and forming the first human society took too long for it to come to pass as a real civilization. Therefore, using Occam's razor principle, the formation of human civilization must be due to an external agent. Anatoly Timofeevich Fomenko is a Soviet and Russian mathematician, professor at Moscow State University. He is well known as a topologist, and a full member of the Russian Academy of Sciences. He is a supporter of revising historical chronology. In his book—*History: Fiction or Science? Chronology No. 1, 2, 3* Anatoly Fomenko set up to prove history, especially global history is wrong.

He even goes further to say that basically, everything we are told about history pre-1600 is BS.

Fomenko uses astronomy data to support his argument that history is too long and that many historical events happened more recently than we thought. He claims that astronomy is precise by definition, and a historical dating that can be calculated from information about eclipses should satisfy any researcher. He asks the question: "Which is more reliable - history or hard-boiled scientific facts?" Science cannot afford subjectivity; most of us would feel the same way about history as well. Chronological problems are very serious indeed.

History: Fiction or Science? Chronology, No. 1, 2, 3 *by Anatoly Fomenko*[1]

---

[1] History

Chapter Five
The fates of Human Societies

In the book *Guns, Germs, and Steel: The Fates of Human Societies*, Jared Diamond effectively, coherently, fundamentally, definitively, and entertainingly explains the world's history of the preceding 15,000 years, where the first societies of humans first appeared. He also analyses how the divergent paths taken by these societies had resulted in the pecking order of different cultures in the world today.

He proves that the Eurasian land mass had by far the best biological resources with which to develop agricultural societies, and was thus more able to form large, coherent, and powerful social entities. Certain kinds of plants and animals are necessary to support a farming society. He investigates the biological resources available to potential farmers in all parts of the world. The people of Eurasia had access to a suite of plants and domesticated animals that provided for their needs. Potential farmers in other parts of the world didn't—and so their fertile soil went untilled. After establishing this strong agricultural foundation, the formation of large-scale societies became possible.

His expose is close to the truth. However, Will Hart in his book The Genesis Race argued that:

Our guts are still not adapted to digest uncooked grains. After all, we are not birds.

Our Paleolithic ancestors lacked the technology to harvest, thresh, process, and cook wild grass seeds. The seeds of wild species are miniscule and they are attached to the seed heads, making them difficult to harvest and hardly worth the effort.

If they lacked an extended experience with wild grasses, how did they know which ones to select to turn into wheat, rye, corn, barely, and rice? In other words, these are still the principal food crops that our civilizations are based upon.

After at least 5,000 years of continuous agriculture, we do not seem to have improved upon the first selections of food crops of our scientifically ignorant ancestors. That hardly seems logical.

This amazingly prescient selection of wild seeds seems not only more than a little surprising, it looks to border on being a minor miracle. There are an estimated 195,000 flowering plants that they could have turned into food sources and primitive man chose less than.01 to base agriculture upon.

The earliest human societies happened at a point in time when people had no concept of domesticating plants or animals, which means no experience with artificial selection.

Will Hart concluded that the Anunnakis brought domesticated plants and animals for humans.

Guns, Germs, and Steel: The Fates of Human Societies by Jared Diamond[2]

---

[2] The Genesis Race: Our Extraterrestrial DNA and the True Origins of the Species *by Will Hart*

Part 8
Pseudo-Science, Proto Science, True Science

Chapter One
Alchemy and The Periodic Table

Alchemy

Alchemy, started as a pseudo science, evolved into a protoscience which then displayed the transformation from theoretical dogma to the observation and practice based methods that gradually developed into a true science. In fact, it can be argued that, in terms of the development of science, the alchemists were further ahead than many other disciplines.

In it infant stage, alchemy became notorious for attracting tricksters and fraudsters. Many so-called alchemists claimed that they had produced gold from base metals, as a means to attract sponsorship or court favor, but most used magician's tricks and misdirection to switch base metals with gold. In the dream of finding riches, everyone from popes and kings to peasants practice the art and try to change base metal into gold.

To obtain gold with the most efficient methodology and technique, many later alchemists conducted controlled experiments and used trial and error to discover the nature of substances, studying how they reacted, interacted and changed under new conditions thereby laying the foundations for chemistry. Alchemists correctly observed phenomenon, came very close to the truth with astute observations, but slipped into the occult and spirituality.

After shifting the administration of cures away from herbalism and mysticism, religion, and theology, alchemy began to take a different direction and move towards science. Alchemists generated a hypothesis and tested it; if it did not work, they refined their ideas and retested, a process intimately familiar to most modern scientists.

In time alchemists moved away from uncovering knowledge for the sake of knowledge and the discipline of alchemy became linked to physics, metallurgy, medicine and pharmacology. It became a true science.

Period Table

In 1869, the Russian chemistry professor Dmitri Mendeleev developed the first Period Table. He organized the 63 then known elements into groups with similar properties and left some spaces blank for those whose existence he could not yet prove. Chemists then began filling in the blanks and started the search for missing elements. The Table was later reorganized using atomic number instead of atomic weight by physicist Henry Moseley in 1913. The expansion of the table supports the idea that as-yet-undiscovered stable or artificial elements exist, which in turn supports the expansion of the Period Table. No one knows for sure if the Periodic Table is ever going to be completed.

Conclusion

The story of alchemy and the period table find the same parallel in Darwin's Theory. Sitchin's Theory will prove to be even more challenging and profound.

Chapter Two
Discovery of Neptune and Pluto

It was perturbations in Uranus' motion that led to the discovery of Neptune.

Astronomers had noticed increasing errors in their models for the orbit of Uranus and eventually ruled out many explanations besides an outer planet that could be perturbing its orbit.

In 1846, the planet Neptune was discovered after its existence was predicted.

This precise prediction of the new planet and its location was striking confirmation of the power of Newton's theory of gravitation.

Later, similar calculations on supposed perturbations of the orbits of Uranus and Neptune suggested the presence of yet another planet beyond the orbit of Neptune. Eventually, in 1930, a new planet Pluto was discovered.

We have even more amazing discoveries recently:

A trans-Neptunian object is any object in the Solar System that orbits the Sun at a greater average distance than Neptune.

Sedna

90377 Sedna is a large trans-Neptunian object, which as of 2012 was about three times as far from the Sun as Neptune. For most of its orbit it is even farther from the Sun than at present, with its aphelion estimated at 937 astronomical units, making it one of the most distant known objects in the Solar System other than long-period comets. Sedna's exceptionally long and elongated orbit, taking approximately 11,400 years to complete, and distant point of closest approach to the Sun, at 76 AU, have led to much speculation as to its origin.

Eris, Pluto, Makemake and Haumea

As of November 2009, two hundred of trans-Neptunian objects have their orbits well-enough determined that they have been given a permanent minor planet designation.

The largest known trans-Neptunian objects are Eris and Pluto, followed by Makemake and Haumea. The Kuiper belt, scattered disk, and Oort cloud are three conventional divisions of this volume of space.

Comet Lovejoy

Comet Lovejoy is a long-period comet and Kreutz Sungrazer. It was discovered in November 2011 by Australian amateur astronomer Terry Lovejoy. The comet's perihelion took it through the Sun's corona on 16 December 2011, after which it emerged intact and continued on its orbit to the outer Solar System. In the process, it astounded all Scientists.

Comet 96P/Machholz 1

Comet 96P/Machholz last came to perihelion on July 14, 2012 and will next come to perihelion on October 27, 2017. 96P/Machholz has an estimated radius of around 3.2km.

Machholz 1 is unusual among comets in several respects. Its highly eccentric 5.2 year orbit has the smallest perihelion distance known among numbered/regular short-period comets, bringing it considerably closer to the Sun than the orbit of Mercury.

Conclusion

Sitchin's interpretation may lead us into accidental discovery of something new, not necessarity discovery of Nibiru. He opened our minds to binary star systems, elliptical orbits and Super Earths. Sitchin allowed us to understand Comet Lovejoy. Comet Lovejoy emerged intact from the Sun's corona on 16 December 2011 and continued on its orbit to the outer Solar System astounding all Scientists.

Chapter Three
Faulty Maps to India led Christopher Columbus to the New Continent

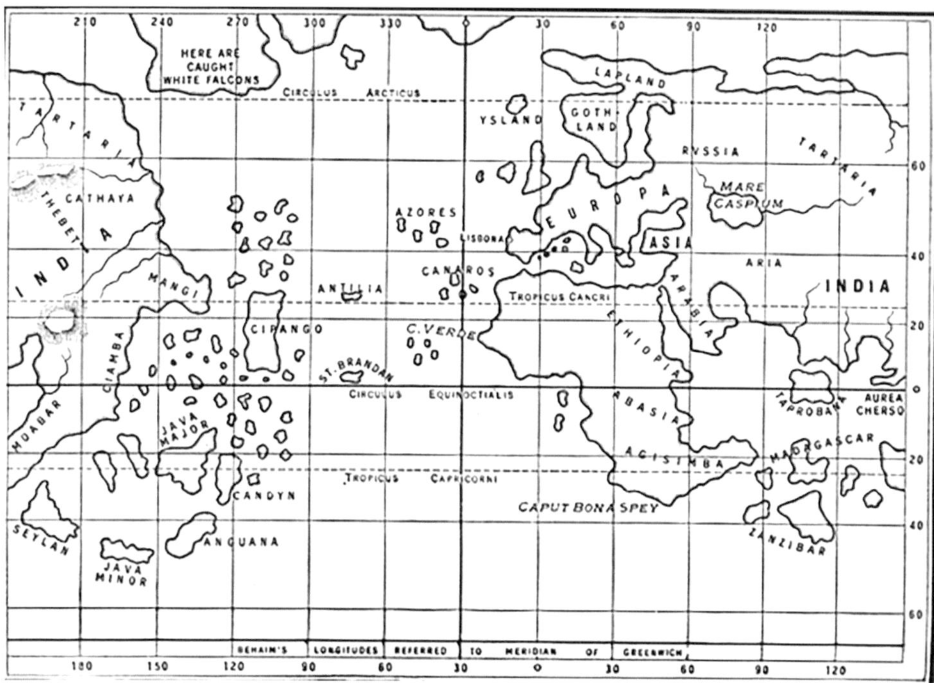

Map used Christopher Columbus 1492

Columbus had dreamed of sailing across the world since he was a young boy, to see if one really would “fall off” the ends of the earth.

During the decade before 1492, as Columbus nursed a growing urge to sail west to the Indies—as the lands of China, Japan and India were then known in Europe—he was studying the old writers to find out what the world and its people were like. He read the Ymago Mundi of Pierre d'Ailly, a French cardinal who wrote in the early 15th century, the travels of Marco Polo and of Sir John Mandeville, Pliny's Natural History and the Historia Rerum Ubique Gestarum of Aeneas Sylvius Piccolomini (Pope Pius II).

Columbus (and most everyone else) already knew the Earth was round. Columbus actually thought the planet was pear-shaped. What was in question, however, was the Earth's circumference. Upon mapping his route, Columbus underestimated the distance to Asia by thousands of miles because he used obsolete Greek data to make his calculations.

He also decided that Spain was closer to China westward than eastward and had never abandoned this conviction.

After years of preparation for his first voyage, Columbus did approach – and was turned down by the kings of Portugal, France, and England for funding.

The discovery of North America was not philosophical in nature at all, because Columbus' intention was to find a faster trade route to China and India, in order acquire more economic status by shortening the route. Finding the Americas was purely an accident.

However, as is frequently noted on this day, when Columbus landed, natives already inhabited the land. He called them Indians because he initially thought he had reached India.

It is written in many historical texts that Columbus and his men may have been responsible for introducing Syphilis to Europe. There is new genetic evidence supporting this theory. Columbus and his men are believed to have brought the disease, which is sexually transmitted, to Europe in 1493. An estimated five million people died in Europe as a result of a massive Syphilis epidemic

Conclusion

If Sitchin's Critics are right, we will have Sitchin looking like Columbus. There is still much to be gained from Sitchin's work.

	Columbus	Sitchin
Guides	Faulty Maps	Missing Tablets
Sources、	Ancient Books	Ancient Texts
Errors	Wrong Distance to Asia	Wrong Interpretation and Translation
Rejected by Mainstream	kings of Portugal, France, and England	Rejected by some Scientists and Historians
Original Purpose	Trade Route to India	Return of Nibiru
Accidental Prize	Finding the Americas was purely an accident	Finding New Planets, Aliens (Nibiru /Annunaki or not)

Chapter Four
Is it better for Galilleo to be born after Newton?

Sir Isaac Newton, born on December 25 1642, was an English physicist, mathematician, astronomer, natural philosopher, alchemist and theologian, who has been considered by many to be the greatest and most influential scientist who ever lived.

Galileo Galilei, born on February 15, 1564, was an Italian scientist who supported Copernicanism, the idea that Earth orbits the sun. Galileo defended his views in Dialogue Concerning the Two Chief World Systems. For doing so, he was tried by the Roman Inquisition, was found "suspect of heresy" and spent the rest of his life under house arrest. His findings changed our world view for all time. Galileo died a natural death in the year 1642. He was 77 years old and was serving his life imprisonment sentence in Arcetri when he died.

In fact, Newton did not discover gravity. It was actually Galileo. Galileo dropped different weight objects from the Tower of Pisa and noted they all fell at the same rate, and did many other experiments along these lines.

In fact, Galileo died a year before Newton was born

What if Galileo Galilei was born after Sir Isaac Newton?

Galileo Galilei would have an easier life, if he was born after Sir Isaac Newton. With the discovery of Gravitational Pull, it could be demonstrated easily that the Earth rotated around the Sun. And he wouldn't spend the rest of his life under house arrest.

Sometimes, some truths or discoveries come too early, especially for people who are indifferent arrogant and ignorant. Politics, greed and an irrational resistance to change have prompted people to denounced the latest discovery or technical innovation as rubbish, often as a knee-jerk reaction. Egos, vested interests, politics, moral and religious objections, and evasion of responsibility are all underlying factors to the fact that many beneficial discoveries had been swept under the carpet.

Many, many lives which can be saved have been lost with the inventor died unheralded. Inventors often need a great deal of tenacity, as well as vision, to overcome the skepticism and ridicule that come as an avalanche. Sometimes discoverers and inventors are prosecuted. A breakthrough would have far-reaching consequences for the entire religious sector based on Galileo Galilei's work. Which, in itself, may be the problem. Until a very long time after his death his books were banned and his writings and findings were considered deviant and false.

Now who is to say what discoveries should come first?

Conclusion

Sitchin may be a Hoax in some people's eyes. We might not value what he has to offered us now. In time when other scientific discoveries are made, they may exonerate Sitchin. To Sitchin's critics, who would like to see his books banned and his writings and findings considered deviant and false, they really need to appreciate how Sitchin's findings have changed our world view for all time.

Chapter Five
Questions for Roman Citizens

"You called yourself the Greatest Empire in existence. Do you believe there exist an Empire called the Chinese Empire which is at least as large if not bigger than The Roman Empire ?"

"You are most proud of your architectural achievements. Do you know there are Pyramids which are very much harder to construct?"

"Other than the Romans and Barbarians, do you know the Eskimos, American Indians, Australian Aborigines…"

"Do you believe your Roman Scholars and Authorities know most of everything?"

"Should you believe in some strange land, strange people, strange ideas and strange discovery if your Roman Scholars and Authorities told you not to?"

"Do you know Korea ? The country is in existence for thousands of years !"

Well, the last question is too hard. Many Americans do not even know Korea existed until the American Government decided to send troops to Korea. Even though The country is in existence for thousands of years ! With this fact in mind, how are we even sure on what we know ? Or for that matter, what we don't know?

Conclusion

Sitchin was trying hard to tell us something we don't know.

Chapter Six
Humans as hybrids of Homo Erectus and Annunaki

A chimera or chimaera is a single organism (usually an animal) that is composed of two or more different populations of genetically distinct cells that originated from different zygotes involved in sexual reproduction. If the different cells have emerged from the same zygote, the organism is called a mosaic. Chimeras are formed from at least four parent cells (two fertilized eggs or early embryos fused together). Each population of cells keeps its own character and the resulting organism is a mixture of tissues.

Hybrid Animals

Lions and Tigers do not exist in the same area, so it would never ever happen in the wild Ligers and Tigons suffer hundreds of birth defects, and usually die young, and painfully. This just shows they are not meant to be born at all.

The female Tiger who has to give birth to the young Ligers is put at risk to, as the Ligers are bigger than her normal cubs would be, she has to have an operation to take them out.

If Ligers don't die when they are young, as they grow their bones don't stop growing, and they become deformed, they are in pain their whole lives. And if they reach and age where they can breed, they are unable to. They cannot reproduce, so the reason every animal is here, is taken away from it, the opportunity to breed is taken away. So other than show, there is no point to these animals.

Dissociative identity disorder (DID), also known as multiple personality disorder (MPD), is a mental disorder characterized by at least two distinct and relatively enduring identities or dissociated personality states that alternately control a person's behavior, and is accompanied by memory impairment for important information not explained by ordinary forgetfulness. These symptoms are not accounted for by substance abuse, seizures, other medical conditions or imaginative play in children

Conclusion

If we believe Sitchin's Theory that we are genetically engineered hybrids, then at once the questions of humanity's Good Vs Evil, constant conflicting emotional states, neurological disorders, junk DNAs and many human medical curiosities can all be solved.

Chapter Seven
abhorrent types of experimental hominids

Elongated Skulls at the Museo Inka (Incan Museum) in Cusco, Peru Groundbreaking work by specialists like the late Dr. Grover Krantz, Dr. Jeff Meldrum, and many others lends credence to the mass of evidence that includes films, audio recordings, hair and scat samples, foot and body prints, distinctively broken tree limbs and animal bones, and other signs of existence from these hair-covered bipeds.

Hominoids are known to come in 4 basic types defined by size:

(1) Bigfoot/Sasquatch are by far the most famous at 7 to 10 feet tall, weighing 700 to 1000 pounds. They live in the deep, dense montane forests that surround the earth at high elevations in temperate latitudes.

(2) Abominable Snowman/Yeti types are human-sized but far more robust than humans, at 5 to 7 feet tall, weighing 500 to 700 pounds. They seem to live only in valleys in the Himalayan Mountain ranges, a combined area as large as the U. S. They are the most primate-like of the four main types, with similarities to chimps and gorillas in physiology and temperament.e">

(3) Alma/Kaptar types are also human-sized, at 5 to 7 feet tall and weighing 500 to 700 pounds, and also living in dense forests around the world, but at lower elevations than bigfoot/sasquatch types. They are the most human-like of the three large types, and I personally suspect these types will prove to be living Neanderthals.

(4) Agogwe/Sedapa types are pygmy-sized at 3 to 4 feet tall and weighing 200 to 300 pounds, living in the thick sweeps of jungles that surround the earth along its equatorial regions. They seem to live in small groups or "tribes," whereas the other three types are believed to function mostly as small independent family units.

The key point to make about these animals, and about the many other variations of names given to them around the world, is that such names simply would not exist if the creatures they are meant to describe did not exist. Where there is so incredibly much smoke, described in details that correspond astonishingly wherever they are found, surely it means a fire is burning, a fire whose flames will eventually reach and consume those who insist it can't be real.

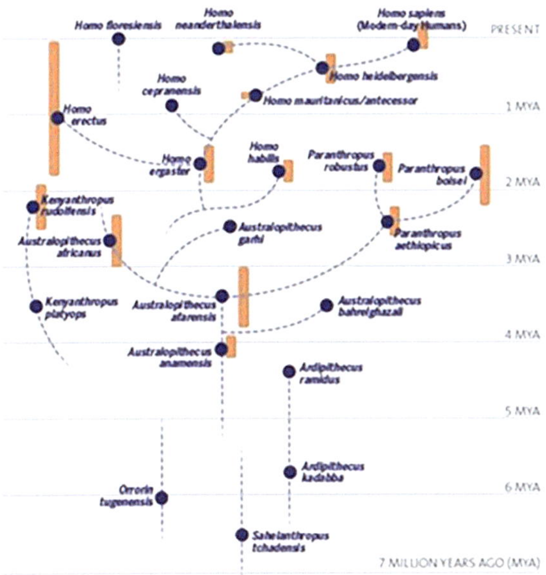

theness.com

Conclusion

Sitchin's Account of Annunaki Genetic Engineering explained the origins of these hominoids.

Chapter Eight

Genesis Revisited: A Scientific Creation Story

December 2001 Michael Shermer

The opening chapters of Genesis will have to be revised to accommodate modern scientific theories and data.

In the beginning — specifically on October 23, 4004 B.C., at noon — out of quantum foam fluctuation God created the Big Bang. The bang was followed by cosmological inflation. God saw that the Big Bang was very big, too big for creatures that could worship him, so He created the earth. And darkness was upon the face of the deep, so He commanded hydrogen atoms (which He created out of Quarks and other subatomic goodies) to fuse and become helium atoms and in the process release energy in the form of light. And the light maker he called the sun, and the process He called fusion. And He saw the light was good because now He could see what he was doing. And the evening and the morning were the first day. And God said, Let there be lots of fusion light makers in the sky. Some of these fusion makers appear to be more than 4,004 light years from Earth. In fact, some of the fusion makers He grouped into collections He called galaxies, and these appeared to be millions and even billions of light years from Earth, so He created "tired light" — light that slows down through space — so that the 4004 B.C. creation myth might be preserved. And created He many wondrous splendors, including Red Giants, White Dwarfs, Quasars, Pulsars, Nova and Supernova, Worm Holes, and even Black Holes out of which nothing can escape. But since God cannot be constrained by nothing (can God make a planet so big that he could not lift it?), He created Hawking radiation through which information can escape from Black Holes. This made God even more tired than tired light, and the evening and the morning were the second day.

And God said, Let the waters under the heavens be gathered together unto one place, and let the continents drift apart by plate tectonics. He decreed sea floor spreading would create zones of emergence, and He caused subduction zones to build mountains and cause earthquakes. In weak points in the crust God created volcanic islands, where the next day He would place organisms that were similar to but different from their relatives on the continents, so that still later created creatures called humans would mistake them for evolved descendants. And in the land God placed fossil fuels, natural gas, and other natural resources for humans to exploit, but not until after Day Six. And the evening and the morning were the third day.

And God saw that the land was lonely, so He created animals bearing their own kind, declaring Thou shalt not evolve into new species, and thy equilibrium shall not be punctuated. And God placed into the land's strata, fossils that appeared older than 4004 B.C. And the sequence resembled descent with modification. And the evening and morning were the fourth day.

And God said, Let the waters bring forth abundantly the moving creatures that hath life, the fishes. And God created great whales whose skeletal structure and physiology were homologous with the land mammals he would create later that day. Since this caused confusion in the valley of the shadow of doubt God brought forth abundantly all creatures, great and small, declaring that microevolution was permitted, but not macroevolution. And God said, "Natura non facit saltum" — Nature shall not make leaps. And the evening and morning were the fifth day.

And God created the pongidids and hominids with 98 percent genetic similarity, naming two of them Adam and Eve, who were anatomically fully modern humans. In the book in which God explained how He did all this, in chapter one He said he created Adam and Eve together out of the dust at the same time, but in chapter two He said He created Adam first, then later created Eve out of one of Adam's ribs. This caused further confusion in the valley of the shadow of doubt, so God created Bible scholars and theologians to argue the point.

And in the ground placed He in abundance teeth, jaws, skulls, and pelvises of transitional fossils from pre-Adamite creatures. One he chose as his special creation He named Lucy. And God realized this was confusing, so he created paleoanthropologists to sort it out. And just as He was finishing up the loose ends of the creation God realized that Adam's immediate descendants who lived as farmers and herders would not understand inflationary cosmology, global general relativity, quantum mechanics, astrophysics, biochemistry, paleontology, population genetics, and evolutionary theory, so He created creation myths. But there were so many creation stories throughout the land that God realized this too was confusing, so he created anthropologists, folklorists, and mythologists to settle the issue.

By now the valley of the shadow of doubt was overrunneth with skepticism, so God became angry, so angry that God lost His temper and cursed the first humans, telling them to go forth and multiply (but not in those words). They took God literally and 6,000 years later there are six billion humans. And the evening and morning were the sixth day.

By now God was tired, so God said, "Thank me its Friday," and He made the weekend. It was a good idea.

Conclusion

It is up to anyone to draw his or her own conclusion.

Chapter Nine
Difficulties on Darwin's Theory

Does Absolute Truth Exist? Are You Absolutely Sure?

The work of the 19th-century English naturalist shocked society and revolutionized science. One of the most influential books in modern history, it has helped shape philosophy, biology, sociology and religion in the 19th, 20th and 21st centuries. But both Darwin's theory and his book

Contain some major flaws.

In his book: The Origin of Species Chapter 6: Difficulties on Theory, Darwin stated:

On the absence or rarity of transitional varieties.

But, as by this theory innumerable transitional forms must have existed, why do we not find them embedded in countless numbers in the crust of the earth?

On the origin and transitions of organic beings with peculiar habits and structure.

It has been asked by the opponents of such views as I hold, how, for instance, a land carnivorous animal could have been converted into one with aquatic habits; for how could the animal in its transitional state have subsisted?

Organs of extreme perfection and complication.

To suppose that the eye, with all its inimitable contrivances for adjusting the focus to different distances, for admitting different amounts of light, and for the correction of spherical and chromatic aberration, could have been formed by natural selection, seems, I freely confess, absurd in the highest possible degree.

Natural Selection is Naturally Wrong

We have eyeballs with retinas and rods and corneas, but so do giant squid.

We do not see fossil examples of animals going from one kind to another, not even one transitional form has ever been found Most of the arguments Darwin used to sustain his theory are rejected by evolutionists of today. The New Evolutionists look at the very same facts and reach radically different conclusions because they bring to the table different lenses through which they interpret those facts in support of Evolution.

Human Evolution is A Theory in Crisis

Darwin claimed that human beings evolved from apes. Today scientists found out that the similarity (between man and chimps) is now down to about 93 percent, according to more recent studies—results that curiously have not made many headlines. Stephan Anitei, science editor for Softpedia, writes: "Well, the new study concludes that the total DNA variation between humans and chimpanzees is rather 6-7%. There are obvious similarities between chimpanzees and humans, but also high differences in body structure, brain, intellect, and behavior, etc." ("How Much DNA Do We Share With Chimps?" Softpedia, Nov. 20, 2006, p. 1).

Sometimes two species are close enough to crossbreed, but the offspring are usually sterile. This is the case when horses and donkeys crossbreed. A male donkey (jackass) and a female horse (mare) will produce a mule. Farmers often preferred mules as work animals prior to the development of the farm tractor. A hinny is the offspring of a female donkey (jenny) and a male horse (stallion). The hinny and mule usually cannot produce offspring. These animals show that Human evolution is not possible.

Is the similarity between chimpanzees and men due to a common ancestor? If that is the case, why are human beings so drastically different now from this ancestor while chimpanzees have remained much the same? The fact is, we are not seeing any evolution presently going on in either chimpanzees or human beings.

As Jonathan Wells notes: "The most fundamental problem of evolution--the origin of species, remains unsolved. Despite centuries of artificial breeding and decades of laboratory experiments, no one has ever observed speciation (the evolution of a species into another species) through variation and selection.

There is no scientific evidence that a species can change the number of chromosomes within the DNA. The chromosome count within each species is fixed. This is the reason a male from one species cannot mate successfully with a female of another species. Man could not evolve from a monkey. Each species is locked into its chromosome count that cannot be changed. If an animal developed an extra chromosome or lost a chromosome because of some deformity, it could not successfully mate. The defect could not be passed along to the next generation. Evolving a new species is scientifically impossible.

Humans have 46 chromosomes. This chromosome count is a steady factor. This determines what is called the "fixity of species" because the chromosome count doesn't vary. People always give birth to people. Dogs always give birth to dogs, etc. The genes can produce variety within the species but cannot result in a different species. Genes allow for people to be short, tall, fat, thin, blond, brunette, etc., but they are still all human beings. The chromosomes make crossing of the species an un-crossable barrier. This certainly would hinder any evolution. Dogs cannot breed with cats. This fact stops evolution dead in its tracks.

Different Perspectives

However, upon closer examination from a different perspective, Darwin's theories has encourage multi-disciplinary scientific progress, problem solving ability and the wide application of some scientific methods.

Progress in many disciplines were made through continuous accumulation and revolutionary processes. Due to his initially outrageous theory, even more brand new ideas are adopted and old ideas become abandoned. In wake of his theory, new methods in science are produced which are used to produce new truths which in turn spawn new scientific theories, which then are used to produce more methods, which are then used to produce more truths in the physical universe. And so on.

His theory reinforce the continuous model of scientific progress that has been set in motion since the Renaissance.

Darwin didn't come up with evolution. What made him special was that he was the first guy to give a mechanism for a unified theory of evolution in which all of life evolved from a simple-cell organism. And the mechanism he gave was this concept of natural selection, an idea that was sparked at the Galapagos Islands.

Charles Darwin was one of those rare individuals who devoted themselves entirely to the pursuit of knowledge. He was, and remains, one of our greatest ever thinkers - the man whose theory of the gradual evolution of living things has changed the way in which most of us see the world in which we live. Because of him we see ourselves as part of nature, instead of separate from it and superior to it.

Evolutionary theory is still such a wonderful and far-reaching view of life despite the flaws and complete body of supporting evidences.

Darwin is still generally recognized as the single greatest thinker in the history of biology, whose contributions provided the basis for understanding the immense diversity that characterizes the natural world.

Darwin's evolution underpins our grasp on how disease-causing organisms evolve and how we can control them with modern medicine. Antibiotic resistance of bacteria and immunity of insects to insecticides are examples of the development of living things by advantageous mutations. HIV – the virus that causes AIDS - naturally mutates to form new strains. So as soon as scientists develop an effective drug that can treat it, a new, resistant form develops.

In the absence of an all-encompassing, multi-disciplinary ideas, Darwin's ideas are still workable ideas of all time in terms of how we understand our relationship with the natural world.

Conclusion

The intelligent Homo Sapiens must be intelligently designed. Many guardians of current Darwinian scientific orthodoxy are casting aspersions to prevent these new insights from Sumerian Clay Tablets from gaining a fair hearing, Whatever teachings, facts and findings Darwinism and Religion have missed or err, Annunaki Intelligent Design could complement and improve upon.

Chapter Ten
Sitchin showed us the way
The Theory of Human Evolution is Wrong, False, and Impossible

New Evolutionists reject old Evolutionists

It is a well-guarded fact that many evolutionists rejected Darwin's theory of evolution over 20 years ago. Stephen Jay Gould, a professor at Harvard University and one of the foremost authorities on evolution in the world said, "The extreme rarity of transitional forms (missing links) in the fossil record persists as the trade secret of paleontologists,...we view our data as so bad that we never see the very process we profess to study". Natural History, Vol. 86. Gould is still an evolutionist, he just rejects much of Darwin's theory.

New Evolutionists reject "Lucy", the Australopithecus

"Lucy", the Australopithecus has been discarded by many evolutionists. The Leakey's considered it simply an extinct ape. It stood three feet tall, had arms that hung down to the ankles and had a brain one third the size of humans. Adrienne Zihlman, U.C. Santa Cruze, said, "Zihlman compares the pigmy chimpanzee to 'Lucy', one of the oldest hominid fossils known, and finds the similarities striking. They are almost identical in body size, in stature and in brain size...indicates that pygmy chimps use their limbs much the same way Lucy did..." Science News, Vol. 123, Feb. 5, 1983, p89

Beneficial Functional Genes can only be added artificially

Scientists today still cannot find any mechanism in nature that adds a gene to benefit the organism. Darwinists claim we evolved from the simplest form of bacterial life to ever more complex forms of life. The most basic bacteria had less than 500 genes; man has over 22 thousand. In order for bacteria to evolve into man, organisms would have to be able to add genes. But there is no genetic mechanism that adds a gene. Mutations change an existing gene but never add a gene. This means there is no mechanism for

Darwinian Human Evolution and this is a fatal flaw in the Theory of Human Evolution."

This is some very strange logic. There is no need for a mechanism to add a gene. What is needed is a mechanism to increase the number of genes. There are several mechanisms for this, and in some ways it happens distressingly often. Naturally, in the great majority of cases this will be a disadvantage, and the organisms affected will not be the most fit and will not replace the 'Normal' type.

This happens in Humans. Down syndrome, which used to be called mongolism, is caused by an extra twenty first chromosome. (In some cases, only part of the extra Chromosome is present.) This occurs in less than 1 in a 1000 in young mothers (less than 30 years old) but much more frequently in older mothers.

There are many other conditions known that are caused by extra genes in Humans. Not all of them are disadvantageous to the same extent.

Chromosome Count Proves Evolution is Wrong.

“DNA is like a computer program, but far more advanced than any software we’ve ever created.” Bill Gates, in the book “The Road Ahead” (Boulder: Blue Penguin, 1996), p228 http://www.arn.org/docs/dewolf/guidebook.htm

There are some serious problems for anyone trying to explain how the vast information system of DNA could spontaneously generate a codex and information encoded using said codex. It is intuitively obvious that codexes exist because intelligences create them in order to record information symbolically. DNA does precisely this and thus at least the DNA molecule must have been designed.

When your dog is going to have a litter, don't worry that she will have a litter of monkeys or cats. She will always have a litter of puppies. The fact that she will have puppies was determined when her chromosomes joined with her mate's chromosomes at conception. You see, a dog has only 22 chromosomes, whereas a monkey has 54 and cats have 38. Half of the total number of chromosomes are contained in the female reproductive cells and half are contained in the male, so the exact total number is brought together in the offspring.

Evidence for Adamu

"By analyzing DNA from people in all regions of the world, Wells has concluded that all humans alive today are descended from a single man also known as Y-chromosomal Adam." Wells wrote the book The Journey of Man: A Genetic Odyssey (2002). http://en.wikipedia.org/wiki/Spencer_Wells

Evidence for Ti-Amat

They "...looked at an international assortment of genes and picked up a trail of DNA that led them to a single (individual) woman from whom we are all descended." "We are finding that humans have very, very shallow genetic roots which go back very recently to one ancestor." Michael Hammer, University of Arizona, Newsweek, 1-11-1988

"Regardless of the cause, evolutionists are most concerned about the effect of a faster mutation rate. For example, researchers have calculated that 'mitochondrial Eve'—the woman whose mtDNA was ancestral to that in all living people—lived 100,000 to 200,000 years ago in Africa. Using the new clock, she would be a mere 6000 years old. No one thinks that's the case..." Ann Gibbons, "Calibrating the Mitochondrial Clock," Science, Jan. 2, 1998, page 28.

Evidence of Creation from a location

"That indicates that there was an origin in a specific location on the globe and then it spread out from there." U.S. News and World Report, 12-4-1995

Conclusion

We have begun the space age. We are all trying to maneuver through a brand new, unfamiliar landscape using an old map drawn up by Religion and Darwin. We need a new guiding principle.

Unaware of the new realities, we are like the last remaining jungle fighters of Japan still hiding in the swamps blindly unaware that the war was over decades ago.

"The past history of Humanity shows us that each stage of its development necessitate an uprooting and renewing of fundamental beliefs in our scientific, social, philosophic and religious conceptions."---M. Plank. Our history is also full of silly mistakes that must make us smile today. Just like new things that happen everyday, another new theory will soon appear, and the new theory that just appeared will be replaced by a newer one, which itself will be replaced by a better one, and so on ad infinitum…Our history is also full of Icons who also made silly mistakes.

Ernest Rutherford declared:' The energy produced by an atom is insignificant.

Those who are expecting to find a new source of energy in this transformation are sweet dreamers". Albert Einstein agreed with him and said :' There is not the slightest indication that nuclear energy will one day be accessible...'Hiroshima happened 12 years later.

Author Arthur C. Clarke said:' When a relatively old, eminent, and distinguished scientist declares that something is possible, he is probably right. But when he says something is impossible, then chances are he is wrong."

Sitchin’s work will make us dream impossible dreams, that we have not dream before. With that, he will bring us to new progress that we can not even dream of, if we understand our past and then foresee our future. It is a new reality too incredible to accept, too awesome to face. The establishment of a new paradigm need not cause the fall of the old paradigm. The time has come to shift into a higher gear and reveal what the future holds for us by examining our very own past.

Neutrons are around for billions of years but we only know they in 1932!!!

Over the years, startling evidence has been uncovered, challenging established notions of the origins of life on Earth—evidence that suggests the existence of an advanced group of extraterrestrials who once inhabited our world.

Many researchers have since uncovered incredible findings having depth, complexity and far-reaching effects in support of Sitchin. This book is an attempt to bring to about the general awareness in the public of a topic so important and so controversial. The reader is encouraged to dig deeper into all the material, the artifacts, the 'what if' and reach his or her own conclusions.

We need to be aware of our own tunnel vision imposed by the current scientific and religious orthodoxy. “Your time is limited, so don’t waste it living someone else's life. Don't be trapped by dogma — which is living with the results of other people's thinking. Don't let the noise of others' opinions drown out your own inner voice. And most important, have the courage to follow your heart and intuition. They somehow already know what you truly want to become. Everything else is secondary." [Stanford commencement speech, June 2005] ---Steve Jobs Wake up! This is not Science Fiction. Truths that were yet unknown to this day have been unveiled.

Part 9
Sitchin Critics

Chapter One
Michael Heiser

Michael Heiser challenges Sitchin's scholarship and translations as spelt out below.

An introduction to Zecharia Sitchin's flawed scholarship and why I feel it is important that these flaws be exposed.

In a nutshell, I'm a trained scholar in Hebrew Bible and ancient Semitic languages and care about my field and its resources. That means I have taken real classes in these languages and the ancient texts from real professors in real universities. I am not stumbling around in the dark. My knowledge isn't just based on the fact that I can use a library. Getting a Ph.D. in this area really does matter. I know many who come to this website are frustrated by "academese" and a seeming unwillingness (it's more than imaginary) of academics to consider alternative research on the ancient world. I would agree with you that there is a "knowledge filter" in academia (I think of Cremo and Thompson's amazing efforts in "Forbidden Archaeology" when I say that), but that does not justify poor scholarship and fabrication of "data" to prop up ideas. It is illegitimate to complain that academics should look at alternate ideas and then turn around and refuse to look at what the original sources say. Whether you want to accept it or not, when you take Sitchin's interpretations of stories over the word meanings the scribes themselves left us (they made dictionaries back then too!), this justifies academics treating alternate material with disdain. This situation should not be. We should look and be willing to slay academic (and even theological) sacred cows; you should respect the results of centuries of work in the field by people who do this for a living.

An analysis of the cylinder seal (VA 243) that Sitchin uses to argue that the Sumerians knew there were 12 planets.

This analysis focuses on the demonstrable fact that the "sun" symbol on this seal (which is essential to allegedly depicting the solar system) is not the sun. The actual sun symbol used on literally hundreds of seals, monuments, and other artwork from Sumer and Mesopotamia is shown to the reader via photos and compared to the symbol on this seal. It's not even close. I include examples where Sitchin's symbol occurs side-by-side with the real sun symbol so there can be no mistaking the fact that the Sumerians and Mesopotamians did in fact distinguish these symbols. This analysis erodes the entire foundation of Sitchin's 12 planet hypothesis.

A study of the word "Nibiru" and an examination of the nature of Nibiru in cuneiform astronomical texts.

The goal here was to amass for readers every occurrence of the word "nibiru" in ancient cuneiform texts. Fortunately, this is possible because of the diligent work of the compilers of the well-known Chicago Assyrian Dictionary, which bases its entries on exhaustive compilations of all cuneiform material known to the present day (there's a reason its taken decades to compile!). The study shows - from the texts themselves, not my opinion - that "Nibiru" is not a planet beyond Pluto and that the Anunnaki gods are never associated with it. These ideas are fabrications. Additionally, this study briefly details the sources left to us by the Mesopotamian scribes that are of an astronomical nature, and addresses Sitchin's "god to planet" matchups that he uses to reconstruct the cosmology of earth and our solar system. In other words, when Sitchin says "the god Marduk is the planet Nibiru" and proceeds to read this equation (and others) into the Sumero-Akkadian texts to interpret them, I compare such equations to the actual lists in cuneiform where Mesopotamian astronomers struck god = planet equations. Not surprisingly, they don't agree.

Zecharia Sitchin's complete misunderstanding of the meaning of the word "elohim."

This study focuses on the fact that, though elohim is morphologically plural (its "shape" or grammatical form is plural), the meaning of the word is almost always singular (one god) in the Hebrew Bible. This is the case over 2500 times. The same phenomenon is also present in Sumerian and Akkadian. The reader does not need to know Hebrew to follow the discussion, as I have color-coded the grammatical features and examples illustrating the truth of this well-known (to those who know Hebrew anyway) feature of biblical Hebrew. The section also contains a response to Erik Parker's (Sitchin's webmaster) attempts to rebut the material. Erik has never studied Hebrew or any ancient language, but he nevertheless tried to respond. It isn't pretty.

Zecharia Sitchin's misunderstanding of the word "nephilim"

This study details the impossibility of Sitchin's translations of "nephilim" as "those who came down" or "people of the fiery rockets" in light of Hebrew vocabulary and grammar. I know it sounds mind-numbing, but again I have tried to illustrate the concepts and problems. It also contains a scan of a page from one of Sitchin's books where he could not tell the difference between Aramaic and Hebrew - an amazing mistake if he's an expert.

Alleged rocket ships in ancient Mesopotamia and the biblical Babel story

The point of this discussion is to show that Sitchin's translations of certain Sumero-Akkadian words cannot be correct for the simple reason that the ancient Mesopotamian dictionaries (yes, they kept bilingual dictionaries and we have them today) translate the words of their own language in ways that unanimously contradict Mr.Sitchin. You either believe him or the ancient Sumerians / Mesopotamians. Seems like an easy call

Chapter Two
Phil Plait

"Planet X doesn't exist, and we are in no danger from a giant planet, rogue or otherwise. The Planet X people are completely wrong. No rogue giant planet is about to destroy the Earth, in May 2003 or otherwise." ----Phil Plait

Ancient texts do not discuss the existence of a tenth planet.

Sitchin's ideas are wrong, and so there is no reason to even introduce the idea of a tenth planet that passes by the Earth.

There is no astronomical indication of the existence of another large planet in the inner solar system.

Planet X has no physical effects, has never been seen, and therefore doesn't exist.

The Sun is not acting in any way abnormally.

The Sun is acting normally, and its behavior does not support the existence of Planet X

There are not more earthquakes than normal.

There no more earthquakes than usual, and therefore earthquakes cannot be used to support the existence of Planet X.

There has been a lot of weird weather lately.

The weather isn't any weirder than it usually is at the end of an El Nino, so the claims of weird weather are wrong.

Brown dwarfs are not at all the way Planet X people describe.

Planet X people do not understand brown dwarfs at all, and saying Planet X is a brown dwarf is actually further evidence that they don't know what they are talking about, and that Planet X doesn't exist.

Observatories are not being closed suspiciously.

Observatories are indeed closing, but only for renovation. Plenty of observatories with public access and large telescopes remain open. Therefore there is no government conspiracy to close telescopes down to prevent people from seeing Planet X.

The pictures that have been posted are either outright fakes or being misinterpreted.

Fake pictures and real ones that are misinterpreted do not support the existence of Planet X.

I am not a government disinformation agent.

There is no evidence that I am anything but who I say I am: a professional astronomer who loves astronomy, and doesn't like hearing the Planet X purveyors misleading others about it. Slurs (and let's face it, slander) against me do not support the existence of Planet X

Chapter Three
Rob Hafernik

Rob Hafernik has the following to comment on Sitchin:

Sitchin's work is a masterpiece of linguistic maneuvering and allegorical interpretation. He is clearly well-read in the archeology and mythology of ancient Sumeria and related lore. His work conflicts, however, with mainstream archeological and scientific opinion. In fact, in Usenet discussions of Sitchin's work, several well-published, respected archeologists familiar with this period have called Sitchin a fool and an idiot (but in less polite terms).

While he's well-read, he seems to live in his own little world when it comes to translation of ancient texts. He also takes certain liberties in his translations and interpretations that are not usually allowed by the scientific community. The part of his work that relates to physics, geology, cosmology and orbital dynamics completely falls apart under the lightest scrutiny. Sitchin's interpretations of Sumerian Epics and other writings describe events that simply couldn't have happened.

Clearly, Sitchin is a smart man. He weaves a complicated tale from the bits and pieces of evidence that survive from ancient Sumeria to the present day. Just as clearly, I think Sitchin is capable of academic transgressions (fracturing quotes, ignoring dissenting facts), "borrowing" of intellectual property and flights of intellectual fancy (the whole book, really). Worst of all, he seems utterly innocent of astronomy and other assorted fields of modern science that are quite germaine to his subject.

He nevertheless paints a picture that is very attractive. One wants to believe it, for it explains so many things. Intellectual honesty, however, prevents anyone with common sense and access to archeological and astronomical data from taking his book seriously. In the end, I think he's just another clever huckster making a living selling books that treat folks to a tale they want to believe in. There are plenty of other sleazy, psuedo-intellectuals out there doing the same thing.

Chapter Four
Ian Lawton

Under Conclusions About Sitchin's Work Ian Lawton in "The Mesopotamia Papers ", Ian Lawton has this to say:

"I have already explained that the reason I have devoted a not insubstantial amount of time and effort to refuting the theories of Zecharia Sitchin is because I believe that, over a number of years, they have misled a great many people about matters of great significance.1 To the extent that, like his former supporter Alan Alford, I was introduced to the enigmas of Ancient Mesopotamia by his work, I do owe him some debt of gratitude. Nevertheless it seems to me a great shame that his ideas are so misplaced that such massive effort is required to correct the balance of opinion in the alternative history community. Were his vivid reconstructions presented in novel form, we could perhaps enjoy them as harmless entertainment. But they are not.

What is my own view of the Mesopotamian texts? I believe that very little, if any, of Sitchin's work deserves to be salvaged. I believe, as I have already hinted on many occasions, that there are certain texts or passages which deserve close scrutiny from an esoteric standpoint; perhaps none more so than the multiple references to the 'creation of mankind'. Although I do not believe the 'gods' were flesh and blood visitors who genetically created man in their own image, nevertheless there are enigmas in these and other aspects of the Mesopotamian texts which are mirrored around the world. However the process of arriving at the most appropriate interpretation thereof is a difficult and lengthy one, not to be undertaken lightly.

However, lest I be accused of continually refuting the theories of others without substituting something positive in return, I can assure my readers that I am currently working on just such a project. I sincerely hope it will be worth the wait..."

Part 10
Sitchin Supporters

Chapter One
Michael Tellinger

Michael Tellinger, author, scientist, explorer, has become a real-life Indiana Jones, making groundbreaking discoveries about ancient vanished civilizations at the southern tip of Africa. His continued efforts and analytical scientific approach have produced stunning new evidence that will force us to rethink our origins and rewrite our history books Michael Tellinger skillfully describes some of the most interesting technologies that the vanished civilizations of southern Africa pioneered:

A vast network of hubs

There are around 10 million ancient stone circles in South Africa, all connected by stone channels, which historians originally thought were designed for cattle. Tellinger suggests these structures were used for gold mining instigated by the Annunaki (a group of Sumerian gods), who created a slave race by genetically tinkering with indigenous proto-humans.

Singing stones

The extremely hard hornfels stone, which Tellinger believes was brought to the area for the express purpose of creating the monuments, has rich tonal qualities and vibrant musicality.

Adam's calendar

This monolithic stone calendar marks time out by the day -- and should be on everyone's bucket list of places to see. Tellinger believes that it's the oldest manmade structure on earth, dating back at least 75,000 years. The calendar, he says, is still accurate today.

What happens in the circle, stays in the circle

Ground-penetrating radar and other tests of the ruins have revealed numerous anomalies, Tellinger reports, such as electromagnetic waves inside the circles and intense heat signatures.

Below are Michael Tellinger's books.

ADAM's
CALENDAR
Discovering The Oldest Man-made Structure On Earth.
75,000 years ago.
By Johan Heine & Michael Tellinger

SLAVE
SPECIES
of god
THE STORY OF HUMANKIND
FROM THE CRADLE OF HUMANKIND
by
MICHAEL TELLINGER

TEMPLES
OF THE AFRICAN GODS
REVEALING THE ANCIENT HIDDEN RUINS
OF SOUTHERN AFRICA

Chapter Two
Marshall Klarfeld

Author Marshall Klarfeld shared his contention that a group of ETs known as the Annunaki visited Earth and genetically altered humans. Basing some of his suppositions on the work of Zecharia Sitchin who translated ancient cuneiform tablets of the Sumerians, Klarfeld said the Annunaki arrived from the rogue planet Niburi, which is on a 3600 year orbit in our solar system. As they needed gold to repair their planet's atmosphere, the Annunaki decided to alter Homo erectus into a new species-- Homo sapiens, so they could serve as miners for them, he explained.

We were "jump started" as a species 250,000 years ago, and are all descendents of the Annunaki, Klarfeld asserted. As evidence of their visitations, he cited he cited various ancient endeavors which he said could not have been constructed by humans alone at that time, including Stonehenge, the huge Baalbek platform in Lebanon, the Giza pyramids, Earth Island statues, the Mitchell-Hedges Crystal Skull, the Nazca Lines, and cylinder seals.

In regards to the Nazca Lines , he shared his new theory: "they are one gigantic time capsule," left more than 3,000 years ago. The figures in the Peruvian desert correspond to the configuration of stars in Orion's belt.

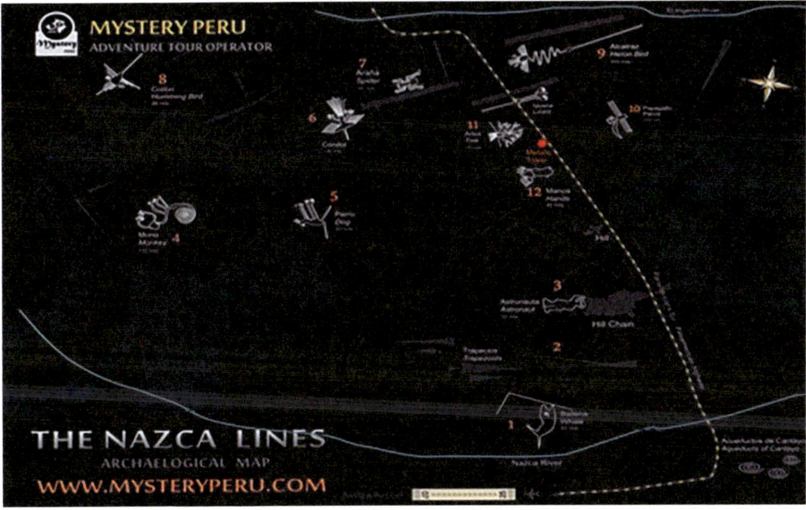

ADAM
the MISSING LINK

The ANUNNAKI Were Here !

Chapter Three
Lloyd Pye

Lloyd Pye has this to say:

In contrast, Interventionists like me anchor our search for origins on evidence rather than faith, on logic rather than magic. We don't think that God did it, or that life spontaneously generated.

For us, evidence and logic point to the same "outside intervention" Intelligent Designers see. However, where they feel the only outside source of intervention must be God (whom they are careful to not mention by name), we suggest another, bolder explanation: "They did it!"

Who are "They"? The currently favored term is Aliens—non-human, non-Earth-based entities.

Of course, aliens raises the hackles and blood pressure of science, government, and religion, so to calm them I will later provide a different, less threatening term. That new term describes entities who have created and distributed, then overseen and managed, life's myriad forms.

Mainstream scientists say extraordinary claims require extraordinary evidence.

Clearly, the Intervention Theory makes several super-extraordinary claims, so we need a great deal of extraordinary evidence. Do we have it?

That human life emerged from primitive hair-covered hominoids (upright walking apes) after human-like entities (aliens or gods, with a small "g") intervened genetically (with test tubes) to create a new hybrid being (humans) with genes from themselves and the primitive hominoids.

The sad truth is that in every field of science, Young Turks have to serve their leaders when those leaders are the oldest, crustiest, and most conservative members of the field. Then, when those old "defenders of the faith" die out, what were once Young Turks take over for them and are forced to defend the same bankrupt faith. "

INTERVENTION
THEORY
ESSENTIALS
Lloyd Pye
EBOOK

EVERYTHING
YOU KNOW
IS WRONG
LLOYD PYE

Chapter Four
Crop Circle Evidences relating to Nibiru

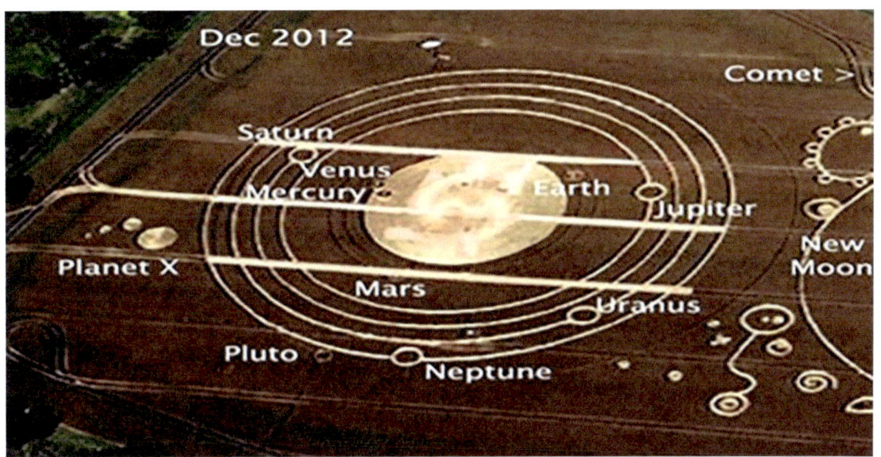

Avebury Manor, England, Jul 22, 2008

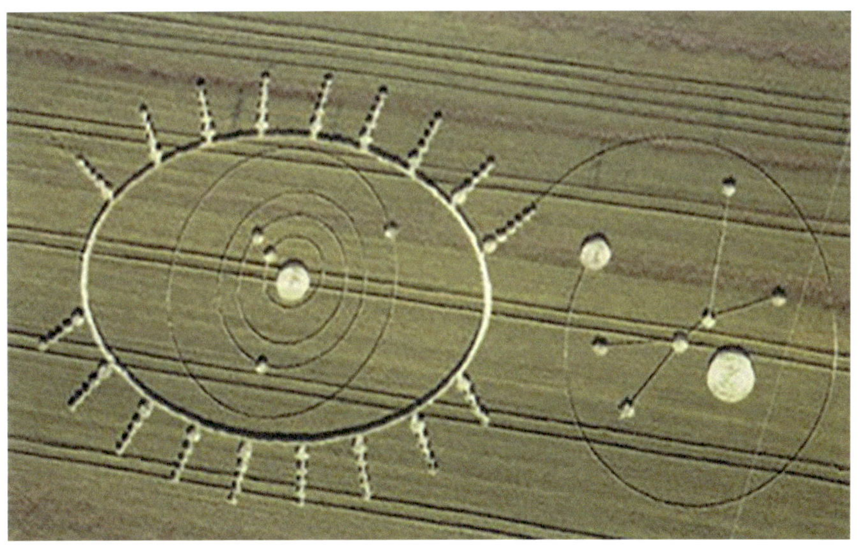

Santena, Poirino. Italy. June 17, 2012

Woodborough Hill. Alton Barnes.Wiltshire. England. Jun 9, 2012

Earth's magnetic field changes into the shape of a "jellyfish" when it is impacted by a severe solar storm

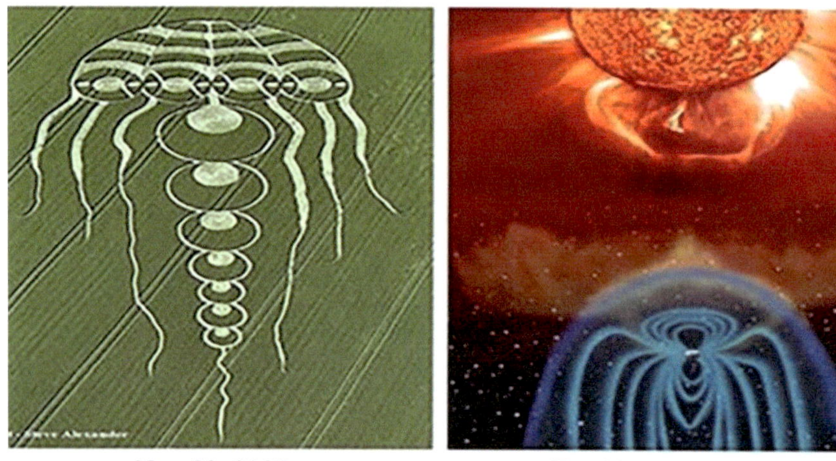

May 29, 2009 **Earth's magnetic field**

600-foot "Jellyfish" crop circle in Kingstone Coombs, Oxfordshire -- June 2009

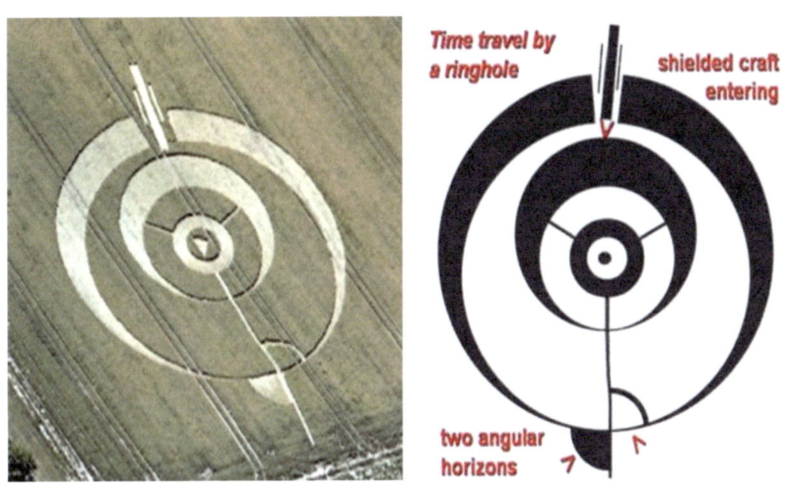

Golden Ball Hill, England, June 29, 2007

Aldbourne, England 'vortex' crop circle, July 2006

Part 11

A brand new interpretation with 21st Century technology vocabularies and scenario descriptions as well as a thorough readaption of the work "The Lost Book of Enki" by Zecharia Sitchin

Chapter One
The debate for Annunaki and Ape Hybrids

B.C. 450,100. In the Court of King Anu, leader of the Annunaki Civilization, which dwelled on the planet Nibiru, the important issue of primitive workers on Colony Earth was raised.

All the elders, wise men, savants, gifted, prodigies, clairvoyant, famous, scientists, leaders, commanders, explorers, entrepreneurs and people's representatives were gathered in the long and bitter related discussions on Morality, Labor Laws, Enslavement Mandates, Interplanetary Journey Stipulations and Planet Nibiru's atmospheric conditions.

The central issue is: Should Primitive Workers be genetically engineered and created to solve the Astronauts Labor Dispute and possible Mutiny?

Major proponents were led by Enki, Second Commander of Terran Operations.

"Make Homo Erectus more intelligent by becoming Homo Sapiens."

"Our commands he will then understand.

Our tools he will then handle.

He shall perform the toil in the excavations.

We will have no more labor disputes coming from our astronauts working as gold miners.

It is not a new creature created by us. The plan is to give this ape creature which we found on Earth more ability, more to our likeness. Only a tiny segment of our DNA is all that is needed."

Opponents were led by Enlil, Supreme Commander of Terran Operations.

"Technological Improvements in Production Methods, Tool Making, Machinary is a better choice. Annunaki should not play God.by creating slave beings!

Creation is the power held by the Father of All Beginnings, not Annunaki

We are sent to Earth to obtain Gold to repair Nibiru's dwindling atmosphere, not to replace the Father of all Beginnings."

There were no verdict after many days of intense debate. On the 7th day, The the Supreme Council of King Anu reconvened for the 7th time.

"Can we get the finest gold dusts in any other way?

Our Planet Nibiru is dying.

The survival of the glorious Annunaki civilization hangs in the balance.

We must get gold to repair Nibiru's broken Ozone Hole.

We must get the finest gold dusts at all costs.

Let's forget about Inter Planetary Travel Rules and save Nibiru.

If there is no Planet Nibiru, there will be no Annunaki Civilization.

If there is no Annunaki civilization, there is no need for Interplanetary Travel Rules.

Let's save Nibiru with creation of slave workers!"

Chapter Two
Story of Garden of Edin, Prehistoric Iraq

Made on Colony Earth: The Finest Gold Powder in the Solar System.

So, the Anunnaki came to the Earth an estimated 450,000 years ago to mine gold in what is now Africa. The main mining centre was in today's Zimbabwe, an area the Sumerians called AB.ZU (deep deposit), The gold mined by the Anunnaki was shipped back to their home planet from bases in the Middle East.

At first the gold mining was done by the intelligent and physically robust Annunaki astronauts. These were men of respectable elite back home. However, they were made to toil and work as miners on Earth in order to save Nibiru. Eventually there were riots and rebellions and the Anunnaki royal leadership decided to create a new slave race to do the work.

Chapter Three
Terran Wild Life Natural Reserve and Research Center
Garden of Edin, Prehistoric Iraq

Enki and his son Ningizidda led a team of biologists to set up a lab for Earth Animal Studies, special interest was given especially to Homo Erectus, a species Enki identified as due to evolve in a few million years into homo sapiens sapiens, the species like the Nibirans.

The apes kept at the Reserve were studied carefully by the Annunaki. Found in the African Forests, the apes walk erect on two legs. Their forelegs they use as arms. They lived among the animals of the Steppe. They eat plants with their mouths. They drink water from lake and ditch. It is a thrill to see them alive! They were kept in strong cages. At the sight of visitors, they jumped up with fists beating on the cage bars, grunting and snorting and speaking no words. Just like the Annunaki, they have males and females!

The study of Homo Erectus revealed that just like the chimpanzees, gorillas and orangutans, they had 24 pairs of chromosomes. Ningishzidda, President of Biogenetics Council explained to Ninmah, the Chief Medical Officer how the second and third chromosomes of Homo Erectus can fused together to give rise to the intelligent New Ape. This fusion could only be accomplished in the Terran Genetic Splicing Laboratory through manipulation of the egg. No disease or natural genetic condition has ever caused a species to alter its chromosomal structure!

(*All members of Hominidae except humans have 24 pairs of chromosomes. Humans have only 23 pairs of chromosomes. Human chromosome 2 is widely accepted to be a result of an end-to-end fusion of two ancestral chromosomes. Wikipedia.)

To the astonished Ninmah, Ningishzidda continued:

"There are things that we can do now, that surpass your wildest dreams. "

"We can jump the gun on the New Ape's progress, and make it into Homo Sapien through infusion of some Annunaki Gene Segements!”

"Oh, Homo Erectus! They are too odd, too monstrous and too barbaric!"

"That is why you are needed for their perfection into Homo Sapien! “

Enki, Second Commander of Terran Operations said to Ninmah, Ningishzidda

"Gather your teams and start working!"

Chapter Four
A.C. T. G. and adenine, cytosine, thymine and guanine
Annunaki Center for Terran Genetics, Garden of Edin, Prehistoric Iraq

Annunaki Human Hybrid Research Commission:

We have the strongest hybrids management systems in the Solar System. Thanks to our xenobiologists' hard work, we now have made most recent accomplishment in Hybrid Gene and DNA formation. Human genetics is an important milestone in our efforts to secure profitable and healthy primitive workers. To make sure that our research and science remain top-notch, new legislation, and King Anu's appropriations for science and management, are critical to maintain this momentum.

Because what we have done so far is historic, we shall commemorate our achievements by encoding the Intelligent Ape's Genetic information as a sequence of nucleotides: adenine, thymine, cytosine and guanine, with The Acronym for our organization A.C.T.G.

Chapter Five
Natural Cycle In vitro fertilisation (IVF)

Admixture of essences was prepared with Mitochondria Extracts, Annunaki Spliced Genes Segments and oval of female Homo Erectus impregnated with Annunaki sperm.

Fertilized oval was inserted back to the Womb of the Ape Homo Erectus.

A birth was expected.

No birth was forthcoming.

In desperation, Ninmah made a dissection and pulled out the baby with tongs.

It was a living being.

Ningishzidda Cried out: "we are successful!"

Ninmah was not filled with joy.

The newborn was shaggy with hair all over.

His foreparts resemble the Homo Erectus.

His hindparts resemble the Annunaki.

The new born was breast fed by the mother Homo Erectus.

The new born was growing fast. A day in the Annunaki Planet is about a month in Africa.

The child grew tall. Not looking like the Annunaki.

His hands could not hold tools.

He could only made grunting sounds instead of speech.

Intravaginal culture (IVC)

Once again, the experiment was done after changing the admixture.

Nimah carefully considered and examined the annunaki essences.

One bit she took from one. One bit she took from another.

In the Nibiru Made Test Tube, the oval of an Earth female was inseminated.

The tube was then hermetically closed and was placed in the maternal vagina and held by a diaphragm for incubation for 44 to 50 hours.

After this time, the content of the tube is examined and embryos are transferred to the uterus.

There was conception.

There was birth giving.

This one look more like the Annunaki.

The baby was breasted by his ape mother and nurtured..

He looked appealing.

Hands fit to hold a tool.

However, he was lacking in senses.

He could not hear and had bad eyesight.

The persistent and patient Ninmah repeated the experiments with varying composition of the admixtures for thousands of times.

Chapter Six

Nimah's Progress Entry

"Hello there, I am Nimah, the Chief Medical Officer for the Terran Genome Project, and I am here to present to you the creature that will change the world as we know it, The Humans.

This creature, can manipulate tools and create different things. For example, give it an object and ask it to turn it into something. It will comply. Most importantly, if you ask it, it will work as miners in our excavations. This creature also holds many wonderful surprises as promised. My team and I will keep trying and trying until the perfect creature as envisioned by our leaders is created."

Hundreds of babies were created each year and their conditions compared.

The dead babies were then dissected and their organs and parts comparatively analyzed

The best case of each year was selected for report. Even then, the best cases were often fraught with defects.

Progress entry 30091 B.C. 450,000: Using Ultra Sound Fetal Scanning, we found this new baby to have paralysed feet.

Progress entry 48762 B.C. 449,999: The New Being had semen dripping.

Progress entry 58973 B.C. 449,998: This new one had Progeria that caused trembling hands.

Progress entry 67554 B.C. 449,997: Ammiocentesis indicated This New One had malfunctioning liver

Progress entry 76545: B.C. 449,996 Upon birth, we found This New One had hands too short to reach the mouth

Progress entry 87656: B.C. 449,995 Through Fetal electrocardiogram (ECG), we could tell This New One had Alveoli Atrophy in the lungs and he could not breath enough air.

Enki was greatly disappointed. "We just cannot create a primitive worker!"

Nimah said to Ningishzidda and Enki, " By trial and error, I am realizing what works."

“We must persist.”

Chapter Seven

Ningishzidda's Junk DNA and Abnormalies Another try. Another deformed creature was made.

It was very late and everyone had left the laboratory except Ningishzidda. In the day time the street was dusty from excavations in the mines, but at night the dust and dew settled. Now at night it was quiet and he could think alone. While he was brilliant in his field, the enigma of life was to him still the greatest mystery. Only the moon kept him company.

Another try. Another deformed creature was made Several weeks later, Ningishzidda again stayed back after everyone had left the laboratory. "Junk DNA sequence." "Maybe Junk DNA sequence is the problem." He kept trying and trying until his eyes could not tell all that junk DNA materials from the junk garbage. And he fell asleep at the table.

Another try. Another deformed creature was made. Like a trial, you don't have to be there, to know the verdict. Ningishzidda knew all the time what it would be. Still, Ningishzidda tried again. Another try. Another deformed creature was made.

Chapter Eight
Curiosity kills the Annunaki

Brainstorming Session after Brainstorming Session was held.

At length, a strange idea came about.

Enki related the fact to Nimah :

"Perhaps the admixture is not wrong. Maybe we need trace elements, nano life substances that were indigenous to earth. They might play a very important role in the functioning and formation of life. We can simulate a life sustaining indigenous admixture from the clay of earth. It would served as a better primordial soup. We should not stick to the conventional thinking of working with test tubes produced by Nibiru Factories. Nor should we set up laboratory conditions similar to Nibiru. We must observe Biological, Chemical and Physical Principles interacting here on Earth."

(Then the LORD God formed man of dust from the ground… Genesis 2:7)

Chapter Nine
FOXP2 language Gene Terran Experiment No. 478,911

Nimah made a vessel using Africa's clay. This she made into a primordial soup.

In the clay vessel, the fertilized Homo Erectus's oval, Annunaki Gene Splices, and Mitochondria Extracts were carefully adusted to right proportions.

Nimah then inserted the fertilized egg back into the female Homo Erectus.

There is conception.

A new born at the awaited time was forthcoming.

It was a male.

It was the image of perfection.

The leaders, Ningishzidda, Ninmah and Enki were overjoyed.

Ningishzidda and Enki were backslapping the baby to make him cry.

Thereby setting the precedent for all human doctors to follow.

The child was growing faster than on their home planet Nibiru.

His limbs were suited for a worker.

However, he knew not speech. He could only make grunts and snorts.

He had the chromosomal disorder FOXP2 and was unable to select and produce the fine movements with the tongue and lips that were necessary to speak clearly.

FOXP2 language Gene was further investigated and then spliced on the Hybrid's chromosome.

Chapter Ten
Frankenstein in Annunaki Womb

Enki's kundalini awakening

Enki reconsidered everything.

Reconsidered everything.

Reconsidered everything...

From Large, structural, chromosomal changes to single nucleotide polymorphisms (SNPs), Enki pondered. Using latest Annunaki Biogenetic Technologies for analysis of genetic variation and genomic profiling, , such as array comparative genomic hybridization, Enki realized that there was one single procedure that was always done. That stood in the way of making the perfect being, with all the attributes that they were determined to achieve.

Into the womb of the Earth Female the fertilized oval was always inserted. "Eureka! This may be the only mistake. We need an Annunaki Gestational Mother. An Annunaki Womb !"

Nimah said to Enki:

"Who would carry a Frankenstein in her womb? "

"Let me persuade Ninki, my wife, " Enki said.

"No, I would dedicate my life in the name of science and face the rewards and endangerment alone." Emphatically said Nimah.

This was the first recorded instance in which the Scientist set the example of using herself as the guinea pig of her experiment.

The fertilized egg was inserted into the womb of Nimah.

There was conception.

Will the conception be nine monthes of Nibiru or Earth ?

It was quicker than on Nibiru and Shorter than on Earth.

He was born perfect.

The newborn uttered the proper sound upon Enki's slapping of its hind parts.

It was the image of perfection.

It was the image of the Annunaki.

(And God said, Let us make man in our image, after our likeness…Genesis 1:26)

Chapter Eleven
Human Prototype Adamu 001

Nimah victoriously shouted: “We’ve done it!”

A complete physical and medical examination was taken of Human Prototype Adamu 001.

It revealed the following:

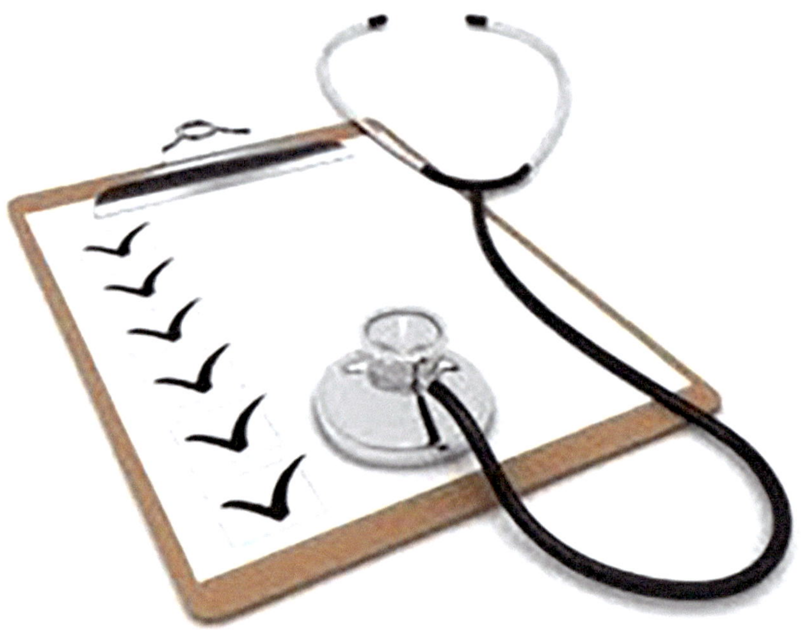

Medical Report Check List Human Prototype Adamu 001

His ear was rightly positioned for hearing.

His eyes was not clogged.

His lower limbs in time will be muscular,

His hands in time can hold tools.

He was not shaggy,

His hair was dark black

His skin was as smooth as Annunaki,

His malehood was surrounded by a foreskin. Unlike that of an Annunaki Male. And this started the tradition of Circumcision in many cultures.

Adamu was truly created in the Image of Annunaki!

Chapter Twelve
Mass Production of Adamu

Female Healers, Nurses, Doctors and Geneticists were taken to Adamu's incubator.

Volunteers were asked to be gestation mothers.

The 7 heroines who step up to the task were remembered as Ninimma, Shuzianna, Ninmada, Ninbara, Ninmug, Musardu and Ningunna.

In the male part of Adamu, Nimah made an incision. A drop of blood to let out.

She squeezed the male part for blood, one drop of blood in each vessel's admixture.

After in situ hybridization, in the wombs of the birth giving heroines, the fertilized ovals were inserted.

At the allocated time, 7 Male Earthlings were born. All were healthy.

7 Primitive Workers have been created.

The Annunakis hope to create the whole race of Primitive workers, eventually.

Chapter Thirteen
Wanted! Gestation Surrogate Mothers!
Sumerian Africa News Network

Sumerian Africa News Network Infomercial:

"There is no greater gift than the gift of life. This is a simple and profound truth.

Very few women are generous, caring and selfless enough to become a surrogate. Will you be one to contribute to the worthy cause of Annunaki?

Please review some basic qualifications and submit a brief application to Annunaki Center for Terran Genetics, Garden of Edin. You will be contacted by an experienced female healer who will explain the process to you and answer any questions."

The Surrogate Mothers Program had proven to be an excruciating task on the Annunaki females.

It was also a huge drain on the resources and manpower.

Ti-Amat, the female companion for Adamu must be fashioned.

So that Adamu and Ti-Amat could reproduce by themselves.

Ninki, Enki's wife was fascinated to create Ti-Amat. After giving her the doctors release and brief her on the dangers, she stood up to the challenge.

Ti-Amat, the first female earthling was born.

Chapter Fourteen
Nature's Curse for Hybrid Animals

Adamus and Ti-Amat were wonders of wonders to hehold in The Terran Wild Life and Natural Reserve Center of Garden of Edin.

When the earthling grew up, there were sexual activity.

But no conception.

This fact had the Annunaki leaders worried a lot.

Ningishzidda order his subordinates to placed hidden cameras in hidden tents by the trees around the cages where the earthlings were lived.

The earthlings behavior patterns were observed and analysed.

Mating, there was.

Conceiving there was not.

Birth giving there was not.

Like hybrid animals, liger, mule and Beefalo, Adamu and Ti-Amat cannot procreate. A natural curse had been placed.

Disappointment engulfed the leaders.

They decided to reexamine Adamus and Ti-Amat 's Chromosomes.

Chapter Fifteen
Genetic Experiment that Goes Horribly Right

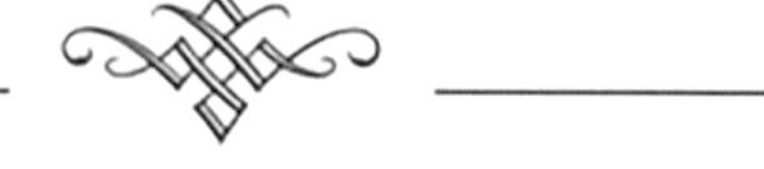

Annunaki Terran Operations Referendum was thrown At the Terran Gene Splicing Quarter.

Every Healer, Geneticist, Doctor, Scientist was invited.

It was of the consensus that in the Human Genome Project, the Annunaki had been too deeply involved.

In the Human Genetic Experiment that Goes Horribly Right, they are giving more and more of their DNA essences to the Humans in the hopes of creating a being that possessed the specific characteristics that they desired, that of a primitive intelligent worker to work in the mines.

It was originally estimated that a drop of their essence would suffice. Now there seemed to be no end to addition of Annunaki Genes...

A great effort was carried to have the sequence of Adamu and Ti-Amat genomes examined, compared and contrasted with Male and Female Annunaki genomes.

It was found that after separating the double helix, there were only 22 pairs of chromosomes. There was no Sex Chromosomes present. And that was nature's curse for hybrid animals.

Chapter Sixteen
ANNUNAKI CLASSIFIED DOCUMENT-- "Genesis"

After the general sessions, behind close doors, Ningishzidda, Ninmah and Enki held a secret discussion. The agenda was put into the classified document---"Genesis", of the highest level. The document detailed the determination of the leaders to go the whole nine yards in creating the perfect being.

The Annunakis had come too far, too involved, and too entrenched to pull out. There was no way to go back. They must make this Human Genome Project a success, even if it means a great chunk of their DNA would have to be given.

In the meantime, everything had to be carefully kept under the dark in case of sure opposition from Supreme Commander Enlil and his supporters who were opposed to the creation of Earthlings as primitive workers from the start. A special task force of Healers, Geneticists, Doctors and Scientist who were sworn to secretcy was thus formed and placed under Ningishzidda.

When the time comes, Ningishzidda had everyone left the Terran Genetic Splicing Lab, except a handful of assistants and himself.

Ningishzidda had induced sleep and applied Anesthesia To Ninmah, Enki, Adamu and Ti-Amat.

From the bone marrow of Enki, he obtained the material for the Y Chromosome.

From teh bone marrow of Ninmah, he obtained the material for the X Chromosome.

In the adult germline stem cells of Admu and Ti-Amat, his secret team of assistants advance their state of the art Annunaki Genetic Biotechnology. After the operations, the wounds in all the ribs of the patients were closed up.

In the Terran Genetic Splicing Lab, with state-of-the-art technology, Patients go home the same day.

(Genesis 2:21-25 And the LORD God caused a deep sleep to fall on Adam, and he slept: and he took one of his ribs...)

Chapter Seventeen
Expulsion from Garden of Edin

Adamu and Ti-Amat were again roaming in the Garden of Edin. They became aware of their sexuality and nakedness. Ti-Amat made aprons from leaves of the wild plants for herself and Amamu.

As Enlil was enjoying his leisure walk in the Garden of Edin, he met the couple. He noticed the aprons on their loins. He was perplexed. And Enki was summarily summoned to explain.

"How do they come to this realization? How do they become so cultured and knowledgeable?

" Now, they must have most of our DNA, and maybe they are also given the gene of longevity!

Like us they shall live for a long long time!"

Ningishzidda and Nimah were also summoned for further clarification. They responded.

" Sex chromosomes were given. The genomic branch for longevity were not given. "

In fits of anger, Enlil expelled Adamu and Ti-Amat from the garden of Edin.

Chapter Eighteen
The Descent of Man

Mass production of prototypes Adamu 001 and Ti-Amat 001 created tribal villages. Sons and daughters were born to Adamu, Ti-Amat and their prototypes. Before long, a population of primitive slave workers were born. With the ability to follow Annunaki commands, the Astronauts from Nibiru were relieved from duty from their gold mine excavations and replaced by Earthlings.

Loads and loads of gold were quickly transported back to Nibiru to make into the finest dust that can be suspended in Nibiru's atmosphere to heal the broken Ozone Layer. The atmosphere was gradually healing. Planet Nibiru was saved and the Great Annunaki Civilization Thrives.

Over time, the children's children of Adamu and Tiamet wandered, mixed and married with of other experimental humanoids thus spoiling Enki's plan of maintaining a quality slave worker race of perfect DNA with perfectly designed attributes.

To counter the bad influence and potential degeneration of population demographics, Enki and his bureau of intellectuals sought to implement the social and hierarchical systems of Annunaki Civilization to the Earthlings.

Chapter Nineteen
Man's First Society

Fauna and Flora were brought from Planet Nibiru and given to the Earthlings as domesticated animals and plants to harvest.

Knowledge of botany, zoology, geography, mathematics, and theology were imparted in schools.

Religions were set up to worship the Annunakis as Gods.

Law codes and social reforms were instituted.

Kingships were established to help rule over the primitive workers.

Summerian language, with precise grammar and rich vocabulary was simplified from the Annunaki Language.

First true cities, having 10,000 to 50,000 inhabitants sprang up.

This sudden civilization which appeared in full swing in Summer, Mesopotamia, today's Iraq over several thousand years ago, became the forbearer of all ancient civilizations, Egyptian, Mayan, Indian, Jewish, Greek, Chinese...

Chapter Twenty
Homo Sapien's final genetic make over

By this time the daughters of man were so civilized, beautiful and sophisticated that the descendants of Annunaki pursue them. Offsprings with extra intelligence, physical strength and extraordinary abilities were born. They became ancient heros across all cultures.

(When man began to multiply on the face of the land and daughters were born to them, the sons of God saw that the daughters of man were attractive. And they took as their wives any they chose.

Genesis 6:1–4)

Chapter Twenty-One
Annunaki Xenobiologists' Science Digest

July 449,500 B.C. Issue

A Study of Hybrid Animal and Homo Sapiens

A study on the hybrid animals of Earth revealed similarities in Homo Sapiens, which were essentially Annunaki and Ape hybrids.

It is extremely rare in the wild and the majority of the hybrid offspring, Ligers (Male Lion and Female Tiger) and Tigons or Tigrons (Male Tigers and Female Lions), are bred in captivity.

The lifespan of ligers, as well as other hybrid animals, is shorter than a normal species. The animals seem prone to cancers and other illnesses. It is possible that the mix of genes contributed to the illness.

The size and appearance depends on which subspecies are bred together. The smaller size of the tigress compared to the lion means that some or all of the cubs may be stillborn or the cubs may be born prematurely [there isn't enough space in the womb for them to develop any further] and may not survive. Premature birth can lead to health problems in those that survive.

Female tigons and ligers have often proved to be fertile and can mate with a lion, tiger or in theory with another species such as leopard or jaguar. Tigons and ligers have been mated together to produce ti-ligers (tig-ligers). Tigers and tigons have been mated to produce ti-tigons. Ti-ligers and ti-tigons are more tigerlike (75% tiger). Ti-tigons resemble golden tigers but with less contrast in their markings. It is possible that the mix of genes contributed to the illness...

Chapter Twenty-Two
Annunaki confronts changing demographics
Sumerian Africa News Network

Annunaki Colony's changing Demographics:

Humans were given the ability to procreate prolifically by Enki and this had led to an explosion in the human population which threatened to swamp the Anunnaki, who were never great in number.

The dilemma, it seemed, was that these demographic changes were on a trajectory that seemed unlikely to change, which spelled a particular problem for the Annunakis in years to come.

The Annunaki had at last awakened to its existential crisis.

The populous hybrid earthlings were never to Supreme Commander Enlil' liking.

From Global Climate Forecast, he knew a big flood was coming.

He decided to let humans perished in the big global flood.

Chapter Twenty-Three
Global Flood for Humanity's destruction

Enki was on human's side and he had Noah in charge of the Prehistoric Svalbard Global Seed Vault Project. To escape Enlil's detection, an ark, called Noah's Ark, was built on the Mountains.

(The ark came to rest on the mountains of Ararat. Genesis 8 :4)

When the time came, the Anunnaki left the planet in flying craft, as an enormous surge of water wiped out much of humanity. There was no doubt that an unimaginable catastrophe, or more likely catastrophes, were visited upon the Earth between approximately 11,000 and 4,000 BC. The geological and biological evidence is overwhelming in its support of the countless stories and traditions which describe such events. They come from Europe, Scandinavia, Russia, Africa, throughout the American continent, Australia, New Zealand, Asia, China, Japan and the Middle East.

The seeds of Humanity was preserved after the global flood.

Enlil was greatly enraged by Enki's double dealing.

As time passed, he regretted his early decision to let humanity perish.

With the ability to quickly multiply, human settlements again reach epic proportions.

Chapter Twenty-Four
Lording Over the Humans

In time, Colonies grew big and spanning many continents on Earth.

The Anunnaki created bloodlines to rule humanity on their behalf and these are the families still in control of the world to this day. Kingship was granted to humanity by the Anunnaki and it was originally known as Anuship after An or Anu, the ruler of the 'gods'.

Eventually Kingships become prevalent all over the globe.

The royal families and aristocracy of Europe, Asia and Middle East are obvious examples of this.

Chapter Twenty-Five
The quest for Global Domination

The two great rivals for all the major issues facing the Annunaki Colonies eventually become two opposing camps, each controlling a hemisphere. Enki, though the first born of Anu, was subordinate to Enlil because of the Anunnaki's obsession with genetic purity. Enlil's mother was the half-sister to Anu and this union passed on the male genes more efficiently than Enki's birth via another mother.

The Anunnaki had many internal conflicts and high-tech wars with each other, as the Enlil and Enki factions fought for control and desired dominance over this planet.

Any group which was so imbalanced as to covet the complete control of the planet will be warring within itself as different factions seek the ultimate control. There was tremendous internal strife, conflict and competition.

Chapter Twenty-Six
Prehistoric WW3 erupted

In 2024BC, the two opposing camps had thermonuclear exchange.

The scenario was summarized below.

"How smitten is the land, its people delivered to the Evil Wind, its stables abandoned, its sheepfolds emptied.

How smitten are the cities, their people piled up as dead corpses, afflicted by the Evil Wind.

How smitten are the fields, their vegetation withered, touched by the Evil Wind.

How smitten are the rivers, nothing swims anymore, pure sparkling waters turned into poison.

In its glorious cities, only the wind howls; death is the only smell.

How smitten is the land, home of gods and men!

On that land a calamity fell, one unknown to man.

A calamity that Mankind had never before seen, one that could not be withstood.

On all the lands, from west to east, a disruptive hand of terror was placed. The gods, in their cities, were helpless as men!"

(Lost book of Enki chapter 1)

For the next 150 years, under Adad's supervision, most of the Nibirans left Earth from Nazca, Peru for Nibiru.

Part 12
Enki's Lessons for Humanity

Chapter One
Enki's open letter to humanity: Avoidance of last World War

You are here because my spirit has selected you to lead others with wisdom, love and compassion. Homo Sapiens is a species with great promise and potential. I have protected you many times in the past, helping you develop and evolve. Nurtured you when you were vulnerable. Now you must rise up or fall. I am telling you something that you may still be reluctant to accept. Your civilization has come to the same juncture that the Annunaki Civilizsation once was.

I am proud of your achievements so far. As a young species, you have come a long way. You have created a world worthy of admiration by many. You have made exponential progress in areas of science. However, you must now try to establish a universal harmonious love for each other and the planet Earth. You must not evolved into a species that is bent on exploitation of other and exploitation of the environment.

The physical manifestations of what is about to happen to this planet—thermonuclear Winter in one hour –are all prelude to a greater tribulation yet to come.

I sincerely hope that you can avoid it. I have inscribed our history onto Sumerian Clay Tablets for you to learn and avoid our mistakes.

Those who do not learn from history, are bound to repeat it in the future. Do not let the last Chapter of human history be written correctly in the ancient dust of Sumer. It is my hope that the histories of the past become prophecies of the future for humanity.

I have faith that you will find a great purpose that sustains and nourishes your spirit in the darkest of times. Accept the weight of the responsibility to yourself and to your species for this grand purpose.

With love,

Enki

Creator of Humanity

Chapter Two
Last World War in Sumer

The Prehistoric Global Nuclear World War which started in Summer also sparked many exchanges in many continents.

In North America, there is evidence to support ancient nuclear war.

"Our research indicates that the entire Great Lakes region (and beyond) was subjected to particle bombardment and a catastrophic nuclear irradiation that produced secondary thermal neutrons from cosmic ray interactions.

The neutrons produced unusually large quantities of 239 Pu and substantially altered the natural uranium abundance ratios (235 U/238 U) in artifacts and in other exposed materials including cherts, sediments, and the entire landscape.

These neutrons necessarily transmuted residual nitrogen (14 N) in the dated charcoals to radiocarbon, thus explaining anomalous dates."

Richard B. Firestone

There are several places on this planet that contain factual and verifiable evidence of an ancient nuclear war and possibly more than one. Perhaps the most unusual recorded evidence comes to us from India in the form of ancient writings, one or two in documents particular known as "Mahabharata" tells the story of a war that took place in that part of the world:

(it was) a single projectile

Charged with all the power of the Universe.

An incandescent column of smoke and flame

As bright as the thousand suns

Rose in all its splendor…

..it was an unknown weapon,

An iron thunderbolt,

A gigantic messenger of death,

Which reduced to ashes

The entire race of the Vrishnis and the Andhakas.

..The corpses were so burned

As to be unrecognizable.

The hair and nails fell out;

Pottery broke without apparent cause,

And the birds turned white.

After a few hours

All foodstuffs were infected…

to escape from this fire

The soldiers threw themselves in streams

To wash themselves and their equipment.

Until the early 1940′ s no one had ever witnessed an atomic blast. No one had ever seen the destruction caused by such a weapon, and most certainly no one could have described the event in any realistic way yet here in these ancient writings from India is a detailed account of just such an event and the results. In the original writing about this war it was said to have taken place in October 5561 BC (7500 years ago).

Albert Einstein said, "I do not know with what weapons WW3 will be fought, but WW4 will be fought with sticks and stones."

Chapter Three
Memoirs of A Cockroach 2024 B.C.

I paused in my tracks.

Most of human populations and their domesticated animals had died.

Finally, our species, the cockroach just might rule planet Earth.

Many survivors of the World War could not find food and would die off.

I knew that I could go on without eating for over a month.

Many more survivors would not survive the radiation.

My ability to survive a nuclear war, long after humans had perished, would now help me rescue my offspring.

Great number of cockroaches had died due to mechanical injuries.

However, with our ability to reproduce quickly, in large amounts, when compared to humans, we would soon quickly roam this Earth again in large numbers.

Chapter Four
We are waiting patiently for you to join us in the wonderful Galactic Journey of Exploration

Our watchers have stayed back to monitor your progress.

We know you are progressing by leaps and bounds.

The majority of us have since moved from the Solar system to other parts of the galaxy where New Colonies with better endowments were set up.

In time you will find our tracks and join us. Our footprints will guide you.

It's a small step for humanity and a giant step for the Annunaki Earthling coalition.

With love,

Enki

Creator of Humanity

Part 13
Conclusion

Unified framework

Wake up! This is not Science Fiction. Truths that were yet unknown to this day have been unveiled.

Knowledge is power, to find overwhelming power, you must be willing to leap over your believes…Carl Sagan

Over the years, startling evidence has been uncovered, challenging established notions of the origins of life on Earth—evidence that suggests the existence of an advanced group of extraterrestrials who once inhabited our world.

Through 'The Lost Book of Enki'", Zacharia Sitchin, a famous Orientalist and Biblical Scholar, has drawn attention to Summeria and the discovered tablets. The tablets were found in what's now modern day Iraq. Famous locations where a lot of tablets in ancient libraries were found: Sippar, Nippur and Nineveh. Sitchin had supported the book with well over 800 Sumerian, Mesopotamian, Akkadian, Babylonian and Assyrian sources. This leads us to the conclusion that human history is far richer and complex than we were ever taught.

The origin of life is controversial because evolutionary theory cannot explain origins and science in general lacks any agreed upon theory for how life could arise from non-life. Also there are some serious problems for anyone trying to explain how the vast information system of DNA could spontaneously generate a codex and information encoded using said codex. It is intuitively obvious that codexes exist because intelligences create them in order to record information symbolically. DNA does precisely this and thus at least the DNA molecule must have been designed.

Many researchers have since uncovered incredible findings having depth, complexity and far-reaching effects in support of Sitchin. This book is an attempt to bring to about the general awareness in the public of a topic so important and so controversial. The reader is encouraged to dig deeper into all the material, the artifacts, the 'what if' and let's reach his or her own conclusions.

Human Evolutionary Genesis theory aims to provide a unified framework that encompasses all theories on the Mysteries of human origins and the Unexpected rise of human civilizations:

1. Evolutionary Theory

2. Out of Africa Theory

4. Intelligent Design

5. Creationism

6. Ancient Astronaut Theory

7. Panspermia.

Its time now for this speculative and controversial theory of **Human Evolutionary Genesis** that challenge established scientific and historical understandings to become the main stream. Transitioning a theory from fringe to mainstream is a complex process, but with substantial evidence and scholarly support, even the most controversial ideas can eventually gain broader acceptance.

Leap over your beliefs

Carl Sagan's *quote, "Knowledge is power; to find overwhelming power, you must be willing to leap over your beliefs,"* highlights the importance of intellectual curiosity and openness. Sagan suggests that true understanding and discovery often require us to challenge our preconceived notions and embrace new ideas, even if they disrupt our established beliefs. Sagan's call to leap over our beliefs encourages an adventurous and open-minded approach to learning, fostering a deeper understanding of the world and our place in it.

The process by which ideas move from fringe science to mainstream acceptance generally follows several stages. This progression can vary depending on the field and the specific nature of the idca, but here are the typical stages:

1. Initial Proposal
2. Preliminary Research
3. Critical Reception
4. Building Evidence
5. Publication and Dissemination
6. Increasing Acceptance
7. Integration into Mainstream
8. Ongoing Debate and Refinement

Understanding this progression can help in critically evaluating how new ideas develop and gain acceptance within the scientific community.

The origin of life

The origin of life is indeed a complex and debated topic. Evolutionary theory primarily addresses the diversification of life once it exists, not how life initially arose from non-living matter, which is the realm of abiogenesis research. The complexity of DNA and its role in encoding biological information does raise intriguing questions about the origins of life. While some argue that DNA's complexity suggests design, others propose naturalistic processes, like prebiotic chemistry and self-replicating molecules, as potential explanations. The debate continues as scientists explore various hypotheses and seek more evidence to unravel the origins of life. Abiogenesis is one theory that suggests life could arise from non-living matter under certain conditions, though it remains a complex and open question. Evolutionary theory provides a framework for understanding how life develops and diversifies, but it does not yet fully explain the origins of life itself. Intelligent Design (ID) posits that some features of the universe and living things are best explained by an intelligent cause. This contrasts with naturalistic explanations offered by evolutionary biology. The scientific consensus supports evolution as the best explanation for the diversity of life, though ID raises questions about the perceived complexity and design in nature.

Alternative human history

The idea that human history is richer and more complex than traditionally taught reflects a growing interest in reexamining historical narratives. Sitchin's theories and other alternative perspectives challenge conventional views, suggesting that our understanding of ancient civilizations and their potential interactions is incomplete. This viewpoint encourages exploring a wider range of historical sources and interpretations, emphasizing that historical knowledge is always evolving as new evidence and perspectives emerge. Zacharia Sitchin's "The Lost Book of Enki" draws on ancient Sumerian and Mesopotamian texts to propose that advanced extraterrestrials influenced early human civilizations. Sitchin's work, including his use of over 800 sources from various ancient cultures, argues that these texts reveal a history of extraterrestrial contact. The ancient tablets from places like Sippar, Nippur, and Nineveh do provide valuable historical insights and offer a fuller understanding of these ancient texts and their implications.

C.S.I. Principles and Methodology

The theory of **Human Evolutionary Genesis** approached the origins of human species as a **"C.S.I. crime scene investigation"**, involving piecing together evidence from various fields to reconstruct the full story. Here's how this investigative process unfold:

1. Analyzing the Scenes

Investigate ancient tools, art, and settlement patterns to understand early human behavior and cultural development. For example, the 200,000-year-old sites in South Africa with circular stone ruins, monoliths, ancient roads, agricultural terraces, and prehistoric mines provide insights into early human activity.

2. Interpreting the Evidence

Ancient Texts, Writing Systems,Inscriptions, Artifacts serve as tangible links to the past, allowing historians and archaeologists to reconstruct and interpret the histories and cultures of ancient Rome, China, and Egypt. It is the same case with the Sumerian Tablets. Sumerian tablets are crucial primary sources for understanding ancient Mesopotamian civilization. Here's an overview of their significance:

A. Historical and Cultural Insight

B. Literary and Mythological Texts

C. Administrative and Legal Documents

D. Scientific and Mathematical Records

E. Preservation and Study

3 Reconstructing the Events

A. Findings and discoveries from South Africa revealed Sumerian descriptions of Abzu, and many other cities as the lands of the First People--including the vast gold-mining operations of the Anunnaki.

B. Numerous ancient maglithic structures, stone tablets, stone monuments and under the ocaean city ruins revealed how these are laid out according to the principles of sacred geometry

C. Many stone calendars, found prevalently at many early human settlements were aligned with the stars. The petroglyphs, carved into the hardest rock, are nearly identical to the hieroglyphs of Sumerian seals.

D. Across the continents, thousands of square miles of continuous settlements and urban centers for ancient civilizations were found.

Unifying fundamental theories

Human Evolutionary Genesis offers a promising avenue for unifying fundamental theories and develps well-supported theories about human origins based on accumulated evidence. This involve integrating new discoveries from various theories and revising previous understandings from various disciplines.

1. Recognizes that new evidences that lead to updates or changes in the dynamics leading to human origins and the rise of human civilization

2. Approaches human origins as an investigative process and emphasizes the importance of evidence, rigorous analysis, and openness to new findings, ensuring a thorough and evolving understanding of our evolutionary genesis history.

3. Encompasses various existing theories and phenomena under a single theoretical umbrella.

Through its comprehensive versatility.

Key driver of exponential progress

Human Evolutionary Genesis hopes to become a key driver of exponential progress in human history through:

1. Catalyst for Innovation

Challenging Norms: Fringe theories often challenge established beliefs and norms, prompting scientists and thinkers to explore new possibilities. This questioning of the status quo can lead to significant breakthroughs.

Example: The early work on electromagnetism by Michael Faraday and James Clerk Maxwell was initially considered fringe, but it paved the way for modern electrical engineering.

2. Stimulating Research

Exploration and Experimentation: Unconventional ideas drive further research and experimentation, leading to new discoveries. This process helps refine and validate these ideas.

Example: The theory of plate tectonics was initially met with skepticism but became widely accepted after extensive research and evidence gathered from various geological studies.

3. Paradigm Shifts

Transforming Understanding: As unconventional ideas gain acceptance, they often lead to paradigm shifts that transform scientific and technological understanding. This can result in entirely new fields of study or technological capabilities.

Example: The acceptance of quantum mechanics revolutionized physics and led to the development of technologies like semiconductors and quantum computing.

4. Exponential Growth

Acceleration of Progress: The integration of new ideas into mainstream thought accelerates technological and scientific progress, often leading to rapid and expansive growth.

Example: The development of the internet, which began as a fringe idea, has transformed communication, information sharing, and virtually every aspect of modern life.

5. Ongoing Evolution

Continuous Innovation: The cycle of fringe theories challenging mainstream beliefs continues to drive innovation and progress. Each new idea or discovery has the potential to lead to further advancements.

Example: The field of artificial intelligence, once considered speculative, is now a major driver of technological development and innovation.

The author hopes that **Human Evolutionary Genesis**, a completely unconventional theory, can ultimately reshape our understanding of ourselves, our civilization and drive the exponential growth of human progress.

Review Requested:

If you loved this book, would you please provide a review at Amazon.com?

You can also reach the author at authorelvis@gmail.com

www.elvisnewman.com

Printed by Libri Plureos GmbH in Hamburg,
Germany